# The Emperor's New Hydrogen Economy

# The Emperor's New Hydrogen Economy

*Darryl McMahon*

iUniverse, Inc.
New York  Lincoln  Shanghai

# The Emperor's New Hydrogen Economy

iUniverse books may be ordered through booksellers or by contacting:

iUniverse
2021 Pine Lake Road, Suite 100
Lincoln, NE 68512
www.iuniverse.com
1-800-Authors (1-800-288-4677)

ISBN-13: 978-0-595-39229-2 (pbk)
ISBN-13: 978-0-595-83619-2 (ebk)
ISBN-10: 0-595-39229-6 (pbk)
ISBN-10: 0-595-83619-4 (ebk)

Printed in the United States of America

In memory of:

Winnifred Bateman Trottier
(1916–2004)

a loving matriarch who knew more about self-sufficiency,
stretching resources and caring for others than I ever will; and,


Brian Cheesman
(1956–2004)

a wonderful, gentle man who continually
encouraged me to tell the real hydrogen story.

# Contents

# *Preface*

Let's be clear from the outset: I have been criticizing the clamour over the hydrogen economy for some time now. Consequently, I have collected a range of critics who have distorted (but not refuted) what I have said and written.

To clarify my argument:

- I am not saying that the hydrogen economy cannot work at a mechanical level.
- I am not saying that current, practical obstacles cannot be overcome.
- I am not saying that we should stop all research and development into improving existing technologies for the production, storage, transport and use of hydrogen.
- I am not saying that hydrogen cannot be a viable fuel in some specific circumstances (e.g., space exploration).

What I am saying is that given the current state of technology and the extremely limited penetration of sustainable energy sources throughout the industrialized world,

**the hydrogen economy is a really, really bad idea.**

It is likely to remain a bad idea for the foreseeable future. We should not deploy significant resources into trying to implement an infrastructure to produce, store, transport, and use hydrogen as a fuel on a pervasive basis, especially at the expense of other sustainable alternatives.

In this book, I will show why the hydrogen economy is a bad idea, today and for the foreseeable future, and provide some promising alternatives.

My critics have also pointed out that I am a lone voice—the large number of people promoting the hydrogen economy can't all be wrong. That will be for you to decide after reading this book. If enough of you think I am right, perhaps we can avoid spending 30 to 50 years and untold financial resources racing to a dead end, and pursue more reasonable objectives instead. Initially, the world thought Columbus and Galileo were wrong.

Many of the key ideas in this book have been posted for several years on my Web site (www.econogics.com), but it is clear that they need to reach a wider audience. The book format provides an opportunity to expand my basic ideas. Books still provide a degree of portability not yet achieved by most web browsing computers. I will endeavour to post updates at the Econogics Web site for this book.

I am tired of people showing me a fuel cell in a lab, a model fuel cell car, or an electrolysis experiment and declaring that this is proof that the hydrogen economy works. This makes as little sense as someone showing me a polished piece of wood, a bit of polished ivory, and showing that plucking a taut string makes a sound, and then declaring that everyone can write and perform piano concertos. Hamsters can be trained to pull small carts. That doesn't mean we should adopt hamster-pulled carts as our predominant means of moving cargo on land, even if we can scale up hamsters to oxen. Little bits of technology, however cute, do not constitute an economically viable solution.

We need to look at the entire energy cycle at an appropriate scale. Then we need to analyze that cycle before making decisions with such vast consequences for our descendants. The decisions we make now may not affect us much, but they will shape the world for our grandchildren.

Initially, I was drawn in by the promise of the hydrogen economy. My experiences include working in the nuclear energy sector, developing electric vehicle technology (and owning several electric vehicles), a degree and work in the field of business administration (including economics and finance), and research on and installation of several sustainable energy technologies. I have seen interest in many of these fields wax and wane, and many a false promise bear bitter fruit. The more I investigated the potential of the hydrogen economy, the more I realized that the proponents were glossing over the hard realities in favour of the superficial appeal.

I will undoubtedly be accused of using selective evidence to support my case. Guilty. The whole point of this book is to get you to consider the conclusion I have reached. The opposing view has plenty of proponents (the hydrogen herd) and has received ample coverage (and acceptance) in the mainstream media. This book presents the evidence, which requires careful consideration. I hope it will help spark the rigorous analysis that is required. I have provided several pages for your notes at the end of this book, so you can record thoughts and questions for your own analysis.

It is possible that I have missed some facts in the overall debate. There are thousands of people working to make the hydrogen economy a reality today. Billions of taxpayer dollars can cover a lot of salaries. I'm working alone, and don't get paid at all for stating the case of the emperor's new clothes. The real question is: have I captured the essential facts and the reality of the big picture?

In the event that this book becomes popular, it will undoubtedly be attacked by those who stand to profit from the current surge of interest in this field. Do not assume that the ferocity and number of attacks means they are correct. Bluster is the standard defense for the indefensible. Examine the responses to see if they present verifiable facts, and not just empty attacks.

In general, the intended audience for this book is the North American consumer and taxpayer. That doesn't mean the information provided here does not apply elsewhere, but I cannot assume that it does. It does not mean the information is not of value to policy makers, but they should already be aware of this information. I am most familiar with the North American environment, and it is probably the economy with the most to gain from what I have written.

This book is divided into two parts. The first part analyses the hydrogen economy. The second part provides realistic, and generally proven, ready-now alternatives that actually could resolve many of the energy-related issues we are facing.

I have made great efforts to ensure that the information in this book is correct as of the time of publication. While the rate of progress in the area of hydrogen research and development is not nearly as rapid as the proponents would have us believe, there may be regular announcements and possibly even real breakthroughs that will supercede some of the information provided here. For the money being spent, I truly hope so.

In closing, I would like to express my thanks to those that helped me bring this book to fruition. My proof-readers, Craig, Marie, and Doreen, spotted the typographical errors that got past the software, and the spots where my brain supplied words when I reviewed the manuscript that weren't on the page. They also helped set the tone for the book, to make technical material accessible to a lay audience. Rocco DiNardo provided valued advice on the manuscript that I have taken to heart. Hakan Falk gave unstintingly of his time and the breadth of his technical knowledge, and encouraged me to proceed. Thanks also to Peter Bursztyn for casting a critical eye over the content, and suggesting valued improvements and information. Finally, my deep appreciation for the work of my editor Gillian Campbell, whose deft touches have added clarity and conciseness to the text, resulting in a much cleaner finished product while preserving the message and voice of the book. Despite the best efforts of those just mentioned, any errors that may have survived are solely my responsibility. I hope you find this volume to be well worth the price of purchase and your time spent reading it.

Darryl McMahon
Ottawa, 2006

# 1

# *Introduction*

2003.01.28/21:01 Eastern Standard Time, The Capitol Building, Washington D.C.

Excerpts from the State of the Union Address:

*"Our third goal is to promote energy independence for our country, while dramatically improving the environment."*

*"I have sent you a comprehensive energy plan to promote energy efficiency and conservation, to develop cleaner technology, and to produce more energy at home. I have sent you Clear Skies legislation that mandates a 70-percent cut in air pollution from power plants over the next 15 years."*

*"I urge you to pass these measures, for the good of both our environment and our economy. Even more, I ask you to take a crucial step and protect our environment in ways that generations before us could not have imagined."*

*"In this century, the greatest environmental progress will come about not through endless lawsuits or command-and-control regulations, but through technology and innovation. Tonight I'm proposing $1.2 billion in research funding so that America can lead the world in developing clean, hydrogen-powered automobiles."*

*"A single chemical reaction between hydrogen and oxygen generates energy, which can be used to power a car—producing only water, not exhaust fumes. With a new national commitment, our scientists and engineers will overcome obstacles to taking these cars from laboratory to showroom, so that the first car driven by a child born today could be powered by hydrogen, and pollution-free."*

*"Join me in this important innovation to make our air significantly cleaner, and our country much less dependent on foreign sources of energy."*[1]

---

1.  United States White House Press Release
    http://www.whitehouse.gov/news/releases/2003/01/20030128-19.html

In this address, George W. Bush, the 43$^{rd}$ President of the United States of America, boldly abandoned several government-sponsored clean-air and fuel-efficiency transportation initiatives. He bet the future of the automobile on one long-shot technology—hydrogen fuel cell road vehicles—and by implication, a pervasive hydrogen production and refuelling infrastructure.

At a time when record federal government deficits were being forecast (even before the invasion and occupation of Iraq), and in spite of being on record as favouring reducing government expenditures, President Bush committed billions of dollars of American taxpayers' money and other considerable resources into making the hydrogen economy a reality by 2019. Since the 2003 State of the Union Address, the U.S. government has increased the amount of this funding from US $1.2 billion to US $1.7 billion.

The Canadian federal government has committed over $200 million of Canadian taxpayers' money to the same goals. Considering announcements by other national governments, the total is well over US $2 billion, and climbing.

Is this an inspired vision comparable to Kennedy's dream of putting a man on the moon before the end of the 1960s, or is this a cynical hoax perpetrated on the taxpayers of the United States and the citizens of the rest of the planet?

There are four key points that concern me about the hydrogen economy, which are neatly counter-pointed by the rhetoric in the 2003 State of the Union address.

1.  The technology required to implement the hydrogen economy on a mass-market scale is not ready. Fundamental research is still required to make it practical and economical.

2.  The hydrogen economy will be neither sustainable nor based on clean energy.

3.  The hydrogen economy cannot lead to energy independence for North America, as it will actually consume more primary energy than current technologies.

4.  The hydrogen economy, even with Herculean efforts and war-scale resources, will not be implemented and working by the time we will need it. We need to select and implement ready-now technologies on a broad scale while we have the luxury of cheap fossil fuels to support the shift.

Clearly, there is a chasm between the picture painted in the 2003 State of the Union Address and the conclusions I have reached. To aid us in reaching a resolution, let us consider the evidence available.

# 2

## *The Allure of Hydrogen*

After the industrialized world sighed a breath of relief that Y2K was nothing more than an excuse for a good party, and not the end of life as we know it, people (and the mass media) focused on other issues again. There were plenty to choose from: famines, wars, disease, and natural disasters. However, these were old news, they did not capture the imagination of corporate America, the mass media, or even the majority of professional do-gooders. No, we hungered for problems that were not shop-worn, that had the patina of novelty, some sex appeal—something that would sell newspapers.

Enter Global Climate Change, previously known as Global Warming, or the Greenhouse Effect. Here was a problem that had it all! Huge scale: it included natural disasters and worries about our ability to maintain our agricultural base, raising the spectre of famines. Even disease gets a nod, as suitable incubation areas for many vector-borne diseases would migrate and expand. Then there was the connection to the fossil fuels culprit as the originator of increased greenhouse gases, the nominal cause of our imminent misfortune, so we can play the guilt card. The fossil fuel concern permitted us to connect to a couple of other big issues: pollution and smog. We worried about their impact on health and the environment (though these seem to get glossed over for the most part), the declining reserves of oil and natural gas, especially the diminishing reserves within the control of industrialized nations. Inevitably, these concerns led us to consider the main sources of this pollution: electrical generation and our use of cars and trucks, always a hot-button issue for vehicle owners.

With so much about our way of life under question, here was a topic that would affect almost everyone that could afford to buy a newspaper.

Let's begin by considering the state of the automobile in North America early in the 21$^{st}$ century. Fuelled by gasoline burned in an Otto cycle internal combustion engine, the car has remained fundamentally unchanged in almost 100 years (other than some cosmetic changes). It conjures a paradox. Although it conveys an image of freedom and independence, and remains the focus of the North American industrial economy; it is also the single largest source of air, water, and noise pollution. Our transportation fuels are increasingly dependent on foreign sources of oil to the extent that it has become a matter of national security. Despite the negatives, we North Americans still love our cars, and have built our society around them for more than fifty years.

While the automobile has its detractors, the love affair between North American consumers and their cars and trucks (increasingly light trucks in the past decade) continues unabated. On the other side of the ledger, the love of profits from automotive sales and related products by corporate America also knows no bounds. In turn, these corporations are a key source of campaign contributions for the two main American political parties.

Politically, messing with either constituency (car owners or car makers) is tantamount to electoral suicide.

The global climate change story was a great success for the mass media. Inevitably, public opinion figured somebody should do something about this problem, which given human nature, meant somebody else should fix it. Generally governments, the grand repositories of our tax monies, were held responsible. By early 2003, weather-related disasters had increased in frequency and magnitude to the point that the major multi-national corporations, especially insurance companies, were publicly complaining about the negative effects. In 2004, the hurricane season produced a record number of tropical storms that did extensive damage to the Caribbean and south-eastern U.S. Japan suffered a record number of typhoons. 2005 brought hurricanes Katrina and Rita, with devastation so severe we have not yet arrived at totals for the costs. The weather events record book has been effectively rewritten in the past decade. The 1990s remain the warmest decade on record, and the years 1998, 2002, and 2003 are the hottest in our history.

Other imperatives (e.g., diminishing air quality, water and ground contamination from oil leaks and spills, increasing payments for rising oil imports) also required that some action be taken to reduce pollution, health impacts, and dependence on foreign fossil fuel reserves. The questionable American tactic of invading other countries (like Afghanistan in 2002 and Iraq in 2003) for resources can only go so far, as U.S. military forces are already thinly stretched. With dwindling oil reserves, increasing dependence on foreign oil sources, pollution, and climate change in the spotlight, the gas-guzzling car was becoming the focal point for serious action.

Now, imagine a car that is virtually silent, produces no emissions other than drinking water, boasts a higher operating efficiency than the internal combustion engine, and is fuelled by the most abundant element in the universe. Unlike electric cars based on lead-acid batteries (the previous preferred solution of the zero-emissions proponents, such as the California Air Resources Board), hydrogen vehicles can refuel quickly and have a range in excess of 100 miles between refuelling stops. The stuff of science fiction dreams? No matter, it's made to order for the current political predicament.

Even better, the same fuel can be used in stationary applications. This is especially attractive in congested urban areas where electricity demand continues to rise, but high-tension electrical transmission capacity can't keep up. Electrical generation can be provided in a compact form factor, with no air pollution being produced. Potable water is a usable by-product. This is a very attractive alternative to the primary fossil fuels used to produce electricity today. Coal is used primarily, but natural gas is being used more and more. There are still a few oil-powered units remaining. Nuclear fission generation is also still in use. Coal-fired generation in particular has been linked to acid rain for decades, and no new nuclear plants have been put into operation in the U.S. since the 1970s due to concerns about their operation.

Finally, this wonder-fuel—hydrogen—can be produced from environmentally friendly sources. Thanks to the magic of electrolysis, any source of electricity can be used to convert ordinary water into fuel. Solar (or photovoltaic) panels can be used to make hydrogen, as can wind turbines, hydropower, and geothermal energy.

It's the fuel that the National Aeronautics and Space Administration (NASA) has used in the American space program to provide electricity on-board spacecraft—an energy solution that's out of this world!

It's a public relations dream! The motoring public would love it, because they can continue to drive their cars and trucks as usual, but without the guilt associated with air and water pollution.

The environmentalists would buy in because the only emission is drinkable water, and the primary energy sources can be "green" as well.

It can be produced from local energy sources, reducing, or even eliminating dependence on foreign sources of fossil fuels, thereby improving the balance of trade.

The technology sector would join in because there are millions of dollars of research and development work to be done, and billions of dollars for implementation. That activity will serve as proof that government is acting to solve the problem.

From the perspective of John Q. Public, it's a solution with no downside. It's a political winner with the great majority of voters.

The hydrogen herd coalesced quickly behind its message:

### Hydrogen—the perfect fuel!

To summarize:

- No pollution—the exhaust is clean water
- Made from renewable energy sources
- Not dependent on foreign energy sources
- Hydrogen is the most abundant element on Earth—we'll never run out

The juggernaut was rolling. With more than US $2,000,000,000.00 in government research and development money up for grabs, the number of true believers and spokespeople grew by the day.

How could anyone possibly find anything to criticize in this Utopian energy vision?

# 3

# *Common Ground*

To really understand the premise of this book, we need to establish some basic truths and definitions before we move forward to consider the reality of the proposed hydrogen economy.

## *3.1. Desirable Things*

Inevitably, discussions about energy, environmental protection, and the politics that affect these will involve, implicitly or explicitly, a discussion about values. Let's make it explicit and get it out of the way.

Being alive and healthy is preferable to being sick or dead.

Clean air is preferable to air filled with pollutants, toxins, carcinogens, and irritants.

Clean water that is safe to drink is preferable to water that is polluted, brackish, contaminated, salty, or laden with pollutants, toxins, carcinogens or disease-carrying bacteria, viruses or other pathogens.

Healthy soil is preferable to soil that is contaminated by pollutants, toxins, or carcinogens, or soil that is depleted, or rendered infertile by salinization or other effects.

Air, water, public lands, and other commonly-owned properties should not be treated as dumping grounds simply because they are not privately owned. They are owned by the inhabitants of the planet, and deserve care and respect, if only so they can remain of useful service to us.

## *3.2. Other Ground Rules*

Until proven incorrect, the laws of gravity and thermodynamics are assumed to apply on planet Earth. I have outlined these laws below, as they suit our purposes.

## Law of Gravity

Things that are heavier than air will tend to fall toward the center of the Earth, and it takes work (energy) to overcome the effect of gravity.

## Thermodynamics

The first law of thermodynamics states that energy is conserved. Energy is not created or lost, but simply changes form. For example, an electric resistance heater changes electrical energy into heat energy.

The second law of thermodynamics is entropy. Any conversion of energy between useful forms will incur losses, usually resulting in heat or noise.

There are Zeroeth (thermal equilibrium) and Third (entropy wins) laws of thermodynamics as well, but these are not important in the context of this book.

In short, there are no perpetual motion or over-unity energy devices. In the event that you have developed a device that produces more energy than it consumes, please don't bother to tell me about it. Prove me wrong by becoming unbelievably rich selling energy at prices lower than any competitor possibly can.
Basic mathematics works, e.g., 0.5 x 0.5 = 0.25.
The human species requires the planet Earth and its ecosystem in order to survive, at least for the foreseeable future. The reverse is not true.
The human species strives to continue to survive, today and for future generations.
We have a moral obligation to leave the planet habitable for our descendants.

# 3.3. Definitions

## The Hydrogen Economy

The term *hydrogen economy* typically means a pervasive infrastructure that uses hydrogen as the primary fuel and energy storage medium. In an energy-intensive society like ours, this energy medium becomes an effective currency of exchange. Usually, the primary sources of energy to be used to produce hydrogen are not defined.

## Hydricity

*Hydricity* refers to an energy structure that is based primarily on electricity and hydrogen, and where the two can be used somewhat interchangeably due to the wide availability of electrolysis (which converts electricity to hydrogen) and fuel cells (which converts hydrogen to electricity). Hydricity usually infers a reliance on nuclear fission power sources as a primary source for electricity production.

## Fossil Fuels

*Fossil fuels* are those substances extracted from the earth, and believed to be finite in human terms, which are used to provide power or energy, either in raw or refined forms. These include oil, bitumen, natural gas, and coal. They may be used in any number of forms, e.g., gasoline, heating oil, diesel fuel, jet fuel, kerosene, methanol, coal, coal oil, propane, naptha, white gas, butane, etc.

Fossil fuels are a finite resource because we are consuming them far more rapidly than they are being created.

## Sustainable Development

Sustainable development has been described as "development that meets the needs of the people today without compromising the ability of future generations to meet their own needs."[1]

Personally, I prefer the old saying: "We did not inherit the earth from our parents; we are borrowing it from our children." So far, I don't think the next generation is going to be impressed with the quality of our stewardship.

## Sustainable Energy

I expect there are those who may make some semantic quibbles about differences between renewable sources and sustainable sources, or whether any source is truly infinitely renewable or sustainable. For my purposes, when used in the area of energy sources, the terms sustainable and renewable are interchangeable. I mean that if we extract some measure of energy from a sustainable source, there is no significant effect on the materials impacted. I am sure there are limits on how much energy can be extracted from a sustainable source before there is an impact. For example, if a wind turbine interacts with the wind to extract mechanical energy (which may in turn be converted to electrical energy, and again into

---

1.    The World Commission on Environment and Development, *Our Common Future*, Oxford University Press, 1987 (The Brundtland Commission)

mechanical energy or heat), the wind itself must be less energetic than it was before the interaction, at some theoretical level. The relevant question is whether the wind or things dependent on the wind are noticeably impacted.

I also doubt that any energy source is infinitely sustainable. The Sun will cease to shine some day. So a more practical definition appears to be in order. The Earth is believed to be over 4 billion years old. Mammalian life on this planet is believed to go back approximately 200 million years. Primates appeared about 50 million years ago. For the purposes of this book, I think the time-scale of 50 million years is an acceptable (albeit arbitrary) one for setting a working definition for indefinite period of time. So, I'm using 50 million years as my yardstick for sustainable. If an energy source can reasonably be expected to be available in approximately the same quantities in 50 million years from now as it is today, that energy source is sustainable.

Astronomers predict the Sun will shine at something like its current intensity for another 5 billion years, or 100 times longer than necessary to meet my standard. So, for my purposes, energy sources that derive reasonably directly from the power of the Sun (e.g., wind, hydro, solar) are sustainable energy sources. Similarly, I have no reason to believe that the mechanism that provides geothermal heating will end within the next 50 million years (approximately 1% of the age of the planet).

Fossil fuels require considerably more than just sunlight to form. Vegetable matter must be created, and then trapped before it can decompose anaerobically. After that, these deposits must be subjected to millions of years of compression under high pressure and elevated temperatures. Deffeyes[2] describes this process quite well. As we are consuming these resources faster than we are recovering them, let alone forming them, they are not sustainable. There is no reason to believe that there are any fossil reserves of hydrocarbons of any kind on the planet that can last for 50 million years while being consumed at anything like current rates. In fact, finding enough for the next 50 years appears to be an insurmountable challenge.

For more information, you can refer to a set of conversion factors (Appendix B) and a glossary (Appendix C).

---

2.  *Hubbert's Peak*, Deffeyes, Kenneth S., especially chapters 2 and 3

# 4

# *Hydrogen Reality*

Carefully obscured from view is a dark side to hydrogen and the hydrogen economy that the hydrogen herd is careful to ignore or minimize. If we are to make intelligent decisions as a society, we have to be informed about both sides of the story. In this chapter, we explore the aspects of hydrogen that the hydrogen advocates are not bringing to the forefront of the discussion.

## *4.1. There Are No Hydrogen Wells*

For all intents and purposes, hydrogen does not exist in a free form in nature. This is because it is a highly reactive element. So, if we want hydrogen, we have to produce it by breaking up some other compound.

## *4.2. Nature Abhors Free Hydrogen*

Nature abhors free hydrogen almost as much as it abhors a vacuum. Hydrogen wants to react with nearly everything in our atmosphere. For example, it can react with nitrogen, which comprises about 80% of our atmosphere, to create ammonia. Even more so, it can react with oxygen, which makes up a little less than 20% of our atmosphere, to produce water or hydrogen peroxide.

In short, if we want hydrogen to remain hydrogen, we have to store it in such a way that it is completely isolated from the atmosphere. We also must make sure it is isolated from a large number of other common materials, such as steel. This is complicated by the fact that hydrogen is the smallest and lightest molecule we know, so it easily leaks through seals that are airtight or watertight. And once it escapes, it just floats away, because it is lighter than air. If it does not react with anything in the atmosphere on the way up, it will actually just sail up and out of the atmosphere into space, lost forever from the planet. Given the current desire to use hydrogen as a fuel, perhaps it is rather convenient that so much has been bound to other things for the past millennia, so it has not all floated away. Increased usage of hydrogen as an elemental fuel will lead to increased loss of the substance from the planet.

While hydrogen may be the most abundant element on the planet, in reality there is no free hydrogen to be had. However, leakage of hydrogen from storage in large quantities as part of the future hydrogen economy certainly increases the probability of losses of hydrogen from the planet in years to come.

## 4.3. Not New, But Not Ready

Hydrogen has been known since the 1500s, if not earlier. It was used extensively in the 1800s as the primary constituent of city gas—used for municipal lighting and powering some industrial works. It was replaced by a superior technology more than a century ago—electricity.

Hydrogen was used in lighter-than-air aircraft in the first half of the 20th century.

Today there is a mature hydrogen industry that produces, transports, stores, and consumes over 50,000,000,000 kg of hydrogen annually, worth billions of dollars worldwide. It is used in numerous industrial processes, to produce other compounds (e.g., ammonia, fertilizers, etc.). There is established practice based on extensive experience. To date, commercial hydrogen production is based predominantly on fossil fuels, not electrolysis or any other means.

Plenty of research and development is still required to make the hydrogen economy practical, let alone economical. This is the case despite the existence of a hydrogen industry that has existed for over a century. Historically, hydrogen has not been used as an energy carrier on a large scale, but in applications where it makes more sense. Due to its volatile properties, it has not been used as a consumer product.

There are frequent announcements by members of the hydrogen herd of advances in the state of the component technologies, or novel ways of producing hydrogen, or a splashy new prototype hydrogen-powered vehicle. This is all part of the desired noise level intended to provide the illusion of significant progress. It is to be expected as innovators chase after the US $2 billion and growing prize. Some of the announcements are even about real advances, although most are simply re-packaging of old news or misleading hype intended to advance stock prices more than real technological progress.

Millions of cubic feet of hydrogen are produced daily in the U.S. today, before any appreciable demand for powering fuel cells or engines.

Industry has been using significant amounts of hydrogen for over a century. If significant breakthroughs remained for hydrogen, isn't it likely that people working in the hydrogen industry, or the worlds of chemistry, physics or biology, or even the crackpot innovator / inventor would have stumbled upon them by now?

Electrolysis and the fuel cell pre-date alternating current (AC) electricity. Courtesy of the genius of Nikola Tesla, we have the benefits of AC technology. But generation, distribution, and use of electricity have advanced very little since Tesla's time. In reality, when we discover a new technology, the major advances tend to come early in its history, not decades or centuries later. Why would we expect it to be otherwise for hydrogen, which saw extensive use prior to 1900 as a fuel, in the 1930s as a lifting gas, and in the past 50 years as a food-processing additive and in other industrial applications?

Unquestionably, research and development will lead to refinements and improvements in hydrogen technologies. But, it is unlikely that a revolutionary new hydrogen source that is efficient, affordable, sustainable, and environmentally benign will be discovered soon. A market for it has existed for years. The low probability of such a breakthrough occurring within the next decade does not warrant betting the economy, let alone our civilization, on it.

The first World Hydrogen Energy Conference of the modern hydrogen era was held in 1976—some 30 years ago. Succeeding conferences have been held every 2 years to the present time; the 15th was held in 2004.

The hydrogen industry exists on a planetary scale. The research has been ongoing for over a century, and with concentrated effort for over three decades. We are still waiting for the breakthroughs required to make hydrogen economically viable and technically attractive for the mass-market as envisaged by the hydrogen economy.

## 4.4. Standards

So far, hydrogen advocates have not decided whether they want to transport and distribute hydrogen as a (very cold) liquid or as a gas kept at high pressures. In fact, within the first dozen hydrogen filling stations implemented in the U.S., different stations have elected to go different ways on this issue. Within the pressurized gas group, there is not even agreement on a pressure rating. Should it be 200 bar, 350 bar, 700 bar (3,000 psi, 5,000 psi, 10,000 psi) or some other figure to be decided later? When it comes to consumer safety and hydrogen, those are the easy questions.

## 4.5. Infrastructure

While there is an established industry that deals with hydrogen everyday; there is no significant consumer-level retail presence. It is nothing like the countless number of gasoline stations that service today's road-going fleet with liquid fuels. A whole new parallel infrastructure will be required to supply hydrogen fuel to the new fleet of hydrogen-powered vehicles envisioned by the hydrogen herd. However, given the properties of hydrogen (wide explosive range, leakiness, gaseous, lighter-than-air, odourless, colourless), it will not be a simple matter of putting in a new set of pumps beside the diesel fuel, unleaded, and premium. It will require a re-thinking of how to distribute an entirely different type of fuel to consumers now comfortable with self-serve operations. Gas-tight couplings or cryogenically cold fuel will be a real departure from liquid fuels used today. The idea of consumers dealing with cryogenic liquids sends a chill up my spine, if you'll forgive the expression.

There are other hazards associated with hydrogen for consumer use. For example, if the hydrogen is stored at high pressure in a metal tank, and there is a leak (e.g., due to collision or poor maintenance), the static electricity generated by the leak itself is sufficient to ignite the escaping gas.

## 4.6. Cold Weather Impacts

A television commercial that was first aired by General Motors in 2003, shows a car, purportedly powered by a hydrogen fuel cell, dripping a puddle of water onto the road. Consider the implications for this small puddle where significant numbers of these vehicles are operating in climates where the road surface is below the freezing point for water (32°F or 0°C). The ice that would form on the roads would create road conditions that would cause collisions on a massive scale. The alternative would presumably be to store the condensed water on-board the vehicle until it could be drained at a safe location (e.g., a drain located off public roadways). However, this presents a couple of additional issues. This will increase the weight of the vehicle considerably as it travels, as the water produced will weigh nine times as much as the hydrogen consumed. Depending on where the water storage tank is located in the vehicle, it could affect the handling of the vehicle.

Let us assume a conventional gasoline-powered car that carries 50 litres (approximately 14 U.S. gallons) of gasoline in a full tank, weighing 40 kilograms (approximately 90 pounds). Gasoline embodies 8.76 kWh (kilowatt-hours) per litre. Assuming the hydrogen fuel cell electric vehicle (HFCEV) is twice as efficient as the gasoline internal combustion engine (ICE) vehicle, it will require about 220 kWh worth of hydrogen. At 33 kWh/kg, that requires about 7 kg of hydrogen for an equivalent energy storage (in terms of distance traveled between refuelling stops). This will potentially become 63 kilograms (approximately 140 pounds) of water between drainage/refuelling stops. That's about the weight of an additional passenger, and will require a tank larger than 60 litres to hold it.

A quick side note: the distinction between kW vs. kWh seems to confuse many people, and I use both in this book. A kilowatt (kW) is a measure of power—the ability to do work. A kilowatt-hour (kWh) is a measure of energy—how much work you can do. In other words, kW is how much you can lift; kWh is how long you can keep lifting it.

Furthermore, in cold climates, it will be necessary to heat the storage tank to prevent the water from freezing, so that it does not expand (damaging the tank) and can be drained. Providing this heat could potentially consume additional energy, although there will be some waste heat available from the operation of the fuel cell itself during operation, but not while the vehicle is parked.

Of course, some will argue that this isn't a problem worthy of consideration, as hydrogen fuel cell cars won't actually operate in cold climates, due to fundamental limitations of the fuel cells themselves. If so, that severely limits their use as a transportation solution in large parts of North America.

While hydrogen may benefit from storage at very low (cryogenic) temperatures, proton exchange membrane (PEM) fuel cells are not so fond of the cold.

With this in mind, it seems somewhat ironic that so many Canadian companies are at the forefront of fuel cell development.

## 4.7. Emissions—Point of Use

We have already noted that HFCEVs do not have true zero emissions at point of use. They produce significant amounts of water and waste heat. In fact, this combination means that most fuel cells actually produce steam, not liquid water. So, it is necessary to supply condensers to dissipate the heat and cool the water vapour into liquid form. This entails carrying more equipment on-board the vehicle (increasing weight, and thus power consumption by the vehicle) or using more energy to drive the cooling operation, or both. It will not be acceptable to simply vent the steam into the atmosphere, as it poses both a potential health hazard (scalding) and a road hazard (condensing or freezing on the roads, making them more slippery). It is easy to dismiss the potential problems associated with small quantities of steam or water vapour production. However, this is precisely how we reacted to fossil fuel exhaust early in the twentieth century, to our later detriment. Water vapour may seem innocuous, but will it still seem so when being produced from many millions of tailpipes? Remember, there are those who believe water vapour in the atmosphere is a potent contributor to climate change[1].

## *4.8. Emissions—Production*

While hydrogen has the potential to be clean and relatively benign at point of use, it is a dirty fuel when we consider the pollutants and toxins that are by-products of the original production. Most hydrogen today is produced from the steam reforming of fossil fuels; using steam and catalysts to break up natural gas into hydrogen, carbon, and impurities. The other components from those feedstocks (sulphur, benzene, etc) are typically released into the environment as part of the production process. In addition, fossil fuels are burned to produce heat to power the steam reforming process, which creates additional pollution. This is covered in more detail in the chapter "Hydrogen Production".

Let's look at producing and using hydrogen in more detail in the next few chapters.

---

1.   As one example, the American Geophysical Union Web site includes the paper, *WATER VAPOR in the CLIMATE SYSTEM*, dated December 1995, which states: "the most important greenhouse gas is water vapor" http://www.agu.org/sci_soc/mockler.html

# 5

## *Producing Hydrogen*

There are no hydrogen wells.

There are no mines that produce hydrogen.

Hydrogen does not rain down from the clouds.

Hydrogen does not grow in fields, forests, lakes, or oceans.

On Earth, hydrogen is inevitably bound to something else at a molecular level. We have to cleave it from other elements in those compounds (water, ammonia, natural gas, and other hydrocarbons) in order to extract the desired hydrogen. However, this requires a readily available, and presumably economical, energy source.

We all need to recognize this basic fact:

**Hydrogen is not an energy source.**

Hydrogen is an energy storage medium, or energy carrier. In order to make hydrogen, there must always be a real energy source. So, whenever the hydrogen economy proponents confuse this matter (and they often treat hydrogen as if it is an energy source), you must look past the 5,000 pounds per square inch (psi) tank of hydrogen, and ask: "where does the primary energy come from to make that hydrogen?"

In this chapter, we examine some of the current and proposed means for the production of hydrogen. As you might expect with the most common element in the universe, there are dozens of sources and methods for producing hydrogen. This chapter discusses a few possibilities.

# 5.1. Electrolysis

Electrolysis is the process of passing an electrical current (DC, not AC) through a liquid, resulting in a chemical reaction. In our context, it is the means of producing hydrogen by applying an electric current into water via electrodes, causing the water molecules to break up into its constituent atoms: oxygen and hydrogen. As the gases are produced at the electrodes (hydrogen at the cathode or negative electrode; oxygen at the anode or positive electrode), they bubble up out of the water, as they are now considerably lighter than the water. With appropriate construction of the apparatus, the two gases can be collected separately and stored for subsequent use. The energy of the electricity is used to break the molecular bonds, but some is also spent creating heat. The voltage can be quite low. Water can be electrolysed into oxygen and hydrogen with just 1.23 volts of electricity (at 25°C—room temperature).

Distilled water is preferred for the electrolysis process. Impurities in undistilled water may also be electrolysed, but it is possible that they will either plate out on one of the electrodes, or affect the purity of the gases produced. Distilling water typically requires additional energy, in order to turn the water to steam, and subsequently condensing it again in order to collect the purified water.

Given the current means of producing hydrogen, electrolysis is the preferred method for generating hydrogen for feeding to fuel cells. This is because the hydrogen from electrolysis is cleaner than hydrogen produced from fossil sources, and the impurities shorten the life of fuel cells dramatically. Purity is less critical if the hydrogen is to be used in a combustion engine. However, if internal combustion engines are to remain in use, we are more likely to use conventional fossil fuels (e.g., natural gas, gasoline, or diesel fuel) rather than to produce hydrogen from them, for as long as fossil fuels remain affordable.

Electrolysis is the only viable, commercial-scale means of producing hydrogen worthy of consideration as a base for the hydrogen economy, as most of the natural gas (feedstock for steam reforming to make hydrogen) in North America will be gone long before a significant number of HFCEVs ever hit the road. (More on natural gas futures later in the book.)

If hydrogen is to be produced on a larger scale (e.g., for HFCEVs) from electrolysis powered by the existing U.S. grid, greenhouse gases will actually increase. That is the conclusion according to the modelling of Michael Wang of the Center for Transportation Research, Argonne National Laboratories in 2002.[1]

---

1.  *Fuel Choices for Fuel-Cell Vehicles: Well-to-Wheels Energy and Emission Impacts*
    Wang, Michael, Center for Transportation Research, Argonne National Laboratory
    http://www.transportation.anl.gov/pdfs/TA/260.pdf

Electrolysis seems the most likely long-term source for hydrogen production (say thirty to one hundred years from now), given the state of the available technologies today and research on alternatives to date. That is, once the various fossil fuels are too scarce to be used for hydrogen production, and hydrogen fuel cells might be sufficiently prevalent to command a major portion of the market for hydrogen consumption. Which is fine, so long as we have more electricity than we need for other applications. Where is all this electricity supposed to come from? A few suggestions are examined in subsequent chapters.

Commercial electrolysis has been in place for at least 80 years, with multiple vendors of commercial scale equipment (Linde, Electrolyser Corporation, Norsk Hydro Electrolysers, and others). Efficiencies for this equipment are typically quoted as being in the range of 50% to 70%, with efficiency typically rising with the size of the operation. Increased efficiencies are usually the result of operations of increased complexity, using high-pressure equipment and a higher risk associated with the operation. Efficiency is considered to be the energy embodied in the hydrogen produced as a portion of the total energy used to produce that quantity of hydrogen. Statements of efficiencies above 70% for electrolysis typically require some creative accounting, such as assuming that the waste heat can be used beneficially. These figures do not include any efficiency losses incurred in the production of the electricity.

In electrolysis, the energy source is whatever is used to produce the electricity that produces the hydrogen—not hydrogen itself.

## 5.2. Thermolysis

Thermolysis is a process in which water is heated sufficiently to dissociate it into hydrogen and oxygen. The process requires very high temperatures, in the order of 3,300°C (approximately 6,000°F) or higher. This leads to some challenges in the construction of apparatus that can operate at these temperatures, as well as the safety of the operators. There is no commercial application of thermolysis to produce hydrogen today, and no serious research work being done on making it practical.

In thermolysis, whatever produces the heat is the energy source. Hydrogen is not the energy source.

## 5.3. Thermochemical

There are over 100 thermochemical cycles that have been proposed as means of producing hydrogen from water, of which several are currently getting significant attention. In all cases, high temperature reactions (up to 850°C or approximately 1,560°F) including elements other than hydrogen and oxygen are required. In general, the other elements are used as catalysts, and can be recovered for subsequent re-use. To my knowledge, none of the current leading candidates (sulphur-iodine, calcium-bromine, or copper-chlorine) have proven more efficient than electrolysis based on sustainable primary energy sources. Sulphur, bromine, and chlorine are not benign materials for human physiology.

The fundamental process is not without issues. There is a trade-off between the corrosive properties of the process and getting to desired reaction efficiencies. Coming up with membrane materials that can effectively separate the various gases, but also survive for reasonable amounts of time at high reaction temperatures remains a challenge. Keeping the produced gases from recombining upon cooling requires careful design of separation technologies that work at high temperatures.

None of these thermochemical processes are in commercial use today. Most are only at the research and development (R&D) stages, funded by taxpayers. There is no reason to believe that any of these will be appreciably economically or technically superior to simple electrolysis at this stage in their development. Interest in these technologies appears linked to the assumption that there will be a massive build of next-generation nuclear fission reactors in the near future (providing surplus electricity and heated water).

In thermochemical cycles, the energy source is whatever produces the heat that is used to produce the hydrogen. Hydrogen is not the energy source.

## 5.4. Where Will the Energy Come From?

Energy will be required no matter which process one chooses, because hydrogen is not an energy source, only a carrier. Can we produce the energy we will need to sustain the hydrogen economy from sustainable sources?

Have I mentioned that hydrogen is not an energy source?

# 6

# *Hydrogen From Sustainable Sources*

If the hydrogen economy is to be sustainable, clearly the source of the hydrogen must be sustainable.

The line between fossil fuels and sustainable sources is not always sharply delineated. For example, natural gas is clearly a fossil fuel. However, it is primarily methane, with a few impurities. Methane can also be considered a renewable source, as it can be produced from the anaerobic decomposition of organic matter. So, how should one categorize methane? I have elected to treat it as a fossil fuel, as the predominant source of methane used in the industrial and residential sectors comes from the natural gas source. Similarly, as most methanol is derived from fossil fuel sources, I have elected to treat it as a fossil fuel-derived source as well. That is not to say that methane or methanol cannot come from renewable sources (clearly they can), only that for the foreseeable future, if you have a quantity of methane or methanol of unknown origin, probability suggests it would have come from a fossil source.

Proponents of renewable energy like to boast about how fast their utilization is growing. The key thing to remember about renewable energy sources in the United States and Canada, however, is that other than hydropower, their total contribution to our overall energy use is trivial, and not making a significant dent in demand.

According to recent figures from the U.S. Energy Information Agency[1], the total contribution of all renewable energy sources in the U.S. in 2002 was 6.1 quads (quadrillion BTUs), or about 6% of the total energy used. This amount is actually less than the peak of 7.1 quads, achieved in the mid-1990s, and equivalent to the production of renewable sources in 1989. As overall energy consump-

---

1.  *Renewable Energy Trends 2003*, U.S. Energy Information Administration, http://www.eia.doe.gov/cneaf/solar.renewables/page/trends/trends.pdf

tion has risen, the share represented by renewable sources has fallen from 7.5% in 1996 to 6% in 2002.

Figures from elsewhere in the world suggest that most other countries are not doing substantially better. Total contribution from renewable sources in other industrialized countries is in similar proportions overall. The fraction in less developed nations is better, but is primarily based on the use of wood and other organic matter being burned to produce heat.

With the exception of rare countries like Denmark, Germany, Spain, and Iceland, which are leaders in implementing sustainable energy sources, there is no appreciable long-term growth in production of high-quality sustainable energy sources suitable for the production of hydrogen anywhere on the planet. This fact is made more alarming because consumption of finite energy sources continues to increase at an accelerating rate.

Unless the whole world seriously embraces the production of energy from sustainable sources on a massive scale soon, the concept of a sustainable hydrogen economy will be nothing but a hollow joke. The required primary energy sources simply will not exist.

## *6.1. Wind*

Proponents of wind power, such as the American Wind Energy Association (AWEA) have stated that wind is the fastest growing energy source (in terms of year over year percentage growth). This was true while the U.S. government was providing the wind production tax credit (PTC). However, when this incentive was suspended during the first term of the George W. Bush administration (2001-2004), installation of new capacity came to a virtual standstill in the U.S.

However, with the series of large turbines that became available in the late 1990s, the economics of such wind farms in areas with strong wind resources have become reasonably attractive, even compared to subsidized and incumbent energy sources. Large-scale wind power is definitely attractive on an economic and environmental basis. The mythology surrounding current, large wind power turbines related to noise, visual aesthetics, and dramatic numbers of bird kills have been largely debunked now, even if this information has not completely permeated the mainstream media or won over all opposition.

Wind power has one major drawback. It produces energy only where and when the wind is blowing. Of course, this availability is not always synchronized with the places and times that the commercial power grid demands. There are times when base-load generation facilities, notably nuclear fission plants that take a long time to start up and shut down actually produce more power than is required on some segments of the North American grid. At these times, the grid operators do not welcome additional power from wind turbines. These times are typically between 2 AM and 6 AM local time, when demand is at its lowest. Some proponents of wind power are searching for ways to store the energy generated by wind turbines at periods of low demand so that it can be utilized at periods of high demand, known as "peak shaving".

In their quest to overcome this lack of synchronization issue, several authors have postulated wind turbines as optimal energy sources for hydrogen production: myself,[2] Fingersh,[3] Earth Policy Institute,[4] and General Electric,[5] among others.

While the use of wind power to generate hydrogen via electrolysis has a significant intuitive appeal, there are also some disadvantages. The ability to use hydrogen as an energy store does eliminate the issue of wind power producing energy when it may not be required, or conversely not producing energy when it is required. The solution as a whole however, does not meet current real-world needs effectively.

Saddling a wind turbine with a hydrogen generation and storage facility increases the initial capital cost of such projects significantly. The major impediment to increased development of wind power today is the initial capital cost. After all, the fuel is free. Increasing the initial cost of a wind turbine simply makes it harder to justify economically.

In North America as a whole, there is no excess wind power capacity today. Even at base-load (minimal demand) conditions, installed generating capacity from all renewable sources (including conventional hydro) cannot meet this demand. Generation from all renewable sources excluding conventional hydro (which has great value in meeting changing demand over the course of the daily demand cycle) remains trivial relative to demand, even just base-load demand. In short, there is no need for storing energy produced from wind today, or for the foreseeable future—it is all swallowed by the existing market demand.

2.  *Fuel Cell Reality Check*, McMahon, Darryl, EV Circuit magazine (2002),
    http://www.econogics.com/ev/fcevreal.doc
3.  *Optimized Hydrogen and Electricity Generation from Wind*, Fingersh, L.J., U.S.
    National Renewable Energy Laboratory (NREL/TP-500-34364), June 2003
4.  Earth Policy Institute,
    http://www.earth-policy.org/Alerts/Alert14.htm
5.  *Large Scale Wind Hydrogen Systems*, Sept. 2003, Liu, Ellen
    http://www.eere.energy.gov/hydrogenandfuelcells/docs/wind_hydrogen_ge.ppt

The existing grid is more than capable of wheeling power from wind farms when it is available to areas utilizing fossil fuel-based generation at periods of low demand. Current barriers to this approach are not technical, but financial. For example, a utility with a coal-fired plant may prefer to keep it running to meet base-load demand in its area rather than buy power at market rate from a remote wind farm that it does not control. The financial interests of the utility's stock holders do not align with the interests of those that suffer the environmental consequences of running the coal-fired plant in preference to using the power available from the wind farm when it is available. A carbon tax would help to resolve this issue.

It has been suggested that one attractive niche for hydrogen storage from wind would be where wind plants are not connected to the grid. However, it is hard to imagine terrain where it will be easier to install and maintain major roads or a pipeline where a set of transmission lines cannot be installed at least as easily and at lower cost.

While the idea of storing energy from wind power for subsequent transportation and use remains attractive in theory, it will not be a practical reality until we have at least 50, and probably 100, times as much electricity generated from renewable sources (excluding conventional hydro) than is the case today in North America. That is unlikely to happen within 25 years, and probably not within 50 years. Today, wind power accounts for less than 1% of electrical generation in North America, and effectively 0% in other sectors (heating, transportation, etc.). If sustainable energy from wind is to be the primary energy for the clean production of hydrogen on a mass scale, the first step is to invest heavily in wind power production in order to have the necessary electrical generation capacity.

When we do reach a point where there is sufficient generation from renewable sources to justify energy storage on an industrial scale, we will likely look to options more efficient than hydrogen. For example, it has been suggested that we use the turbine towers as hydrogen storage tanks.[6] However, another option would be to use the tower to store ordinary air under pressure, which would also store a significant amount of energy. When power demand exceeds the power being extracted from the wind, the pressurized air can be used to turn generators to produce electricity. This approach uses current technology at significantly lower cost, and is not subject to issues related to hydrogen embrittlement. Pro-

---

6.   *Optimized Hydrogen and Electricity Generation from Wind*, Fingersh, L.J., U.S. National Renewable Energy Laboratory (NREL/TP-500-34364), June 2003, page 4

longed exposure to hydrogen affects the structure of steel, making it brittle and increases the potential for failure.

Perhaps the greatest value that wind generation offers now is as a benchmark. Hydrogen produced from electrolysis using electricity generated by wind power is a sustainable resource, with minimal negative environmental impacts. It also remains an upwardly scalable resource for the foreseeable future that can provide solid financial and physical output numbers. Any other sustainable hydrogen source will need to prove itself cost-competitive with the wind option to be taken seriously.

## 6.2. Photovoltaics

In this book, I use the term photovoltaics (PV) to refer to solar cells that produce electricity. The popular term solar panel creates confusion because it can mean either a device that uses sunlight to warm air, water, or some other substance, or a collection of photovoltaic cells arranged to produce electrical current when exposed to light. Some can produce both electricity and heat. The term photovoltaics allows us to make the distinction between solar cells and other forms of solar energy more easily.

Today, conventional photovoltaic cells typically produce a fairly low voltage (less than a volt). Multiple cells have to be joined in series to produce electrical current at useful voltages, for example, 12 volts DC and up. However, given the low voltage requirement for electrolysis of water to produce hydrogen (as low as 1.23 volts), fewer PV cells would have to be placed into a circuit to drive electrolysis than to charge a conventional storage battery.

The attractiveness of this application palls quickly when we consider a real world application. Taking the electricity produced by the photovoltaic cell as a given (100%), using it to produce hydrogen via conventional electrolysis will incur a 30% to 50% penalty. Using that hydrogen to power a hydrogen fuel cell vehicle will incur at least another 50% penalty. Of the electricity originally produced by the PV cell, only about 30% will be available to the drive train of the vehicle. By comparison, using the same electricity to simply charge the vehicle batteries means that at least 80%, and possibly 90%, of the electricity produced would be available to the drive system of a battery-only electric vehicle.

There has been a lot of discussion suggesting that photovoltaics are net energy sinks (energy consumers), not sources. I guess it comes down to which studies you choose to believe. Assuming that current commercial grade photovoltaics are installed appropriately in locations that receive excellent insolation (energy from sunlight), i.e., more than 6 solar hours per day on average, it is my belief that photovoltaics are net energy sources. Presuming that such panels can deliver their rated output for 20 years or more, they generate more than enough energy to produce replacements, indefinitely.

I cover this issue in more depth later in the book under Personal Energy Production—Photovoltaics.

Clearly, photovoltaics will produce electricity only when the sun is shining on them. There are unquestionably some environmental downsides to photovoltaic production, including extensive water use and toxic residues. For now, the primary issue for photovoltaics is cost. Until photovoltaics can produce electricity at a price competitive with other sources, including other sustainable sources, they are unlikely to be utilized for the production of hydrogen on a commercial scale.

## 6.3. Solar Thermal

Solar thermal covers a range of possibilities from a simple concentrating lens (magnifying glass) that provides more heat at a point than the surrounding area or a solar oven, to grand structures such as solar chimneys, solar furnaces, heliostats, or industrial-scale steam plants that derive their heat from concentrated solar energy.

However, if our objective is to produce hydrogen, we can discount the smaller and simpler contrivances due to efficiencies and economies that come with scale, and focus on the larger (and more complex) structures.

A plant such as the Luz solar electric generating stations (SEGS) built in the California Mojave desert from 1985 to 1991 could be scaled up to produce significant amounts of electricity to electrolyse water into hydrogen and oxygen. The last of the Luz SEGS built could produce 80 megawatts (MW) in full sun. Unlike the original demand load (the California power grid) for the Luz SEGS, production of hydrogen could be set up to produce only when sunlight is available, eliminating the need for dispatchable power. Dispatchable power is a utility term for energy that can be generated when it is demanded, rather than when the sun shines or the wind blows. Furthermore, there is some evidence that electrolysis efficiency can be boosted by heating the water prior to electrolysing it. At a solar thermal plant, the water could be pre-heated using waste heat from the steam-condensing phase of powering an electric generating turbine.

The key to such a facility will be access to quantities of water and a high degree of annual insolation energy. The water need not be perfectly clean. In this environment, it should be fairly easy to steam-distill the water using a combination of the waste heat from the turbine and solar distillation. Situated in the Mojave desert, the various Luz SEGS do not have easy access to large quantities of water. Placement near a lake or river would likely be preferable. A river might permit the building of a combined hydro/solar powered plant, where hydropower could be used to turn the generator shaft in the absence of solar-heated steam. As is the case with photovoltaics, the solar thermal plants will produce energy only when the sun is shining, unless another energy source is also incorporated into the structures.

However, the losses associated with electrolysis and subsequent use in a fuel cell mean that it would be more efficient to feed this sustainable electricity into the grid than using it to produce hydrogen.

## 6.4. Hydro

Hydropower is easily the major source of sustainable electricity generation today. It can also be used to produce mechanical energy, which was more common in various kinds of mills of years past (saw mills, grist mills, textile mills).

Hydro generation has another advantage in that its reservoirs provide a means of storing energy, which is uncommon with most renewable energy technologies (e.g., wind, solar). This storage, and the quick spin-up times associated with water-driven electric turbines means power production from hydro facilities can be brought on-line and off-line very quickly to meet changing demand on the grid. For example, it takes seconds to minutes compared to thermal generating stations that take minutes to hours, and nuclear plants that take hours to days to come on-line.

However, these very advantages make hydro generation a poor choice for producing the electricity to drive electrolysis to make hydrogen. Hydro facilities are too valuable in stabilizing the electrical grid to divert their generating and storage capacity to the production of hydrogen—a less efficient means of storing electrical energy.

While small hydro installations (e.g., low-head, run-of-river, and micro-hydro) have environmental advantages, their small size disqualifies them from producing electricity to produce hydrogen. Output measured in watts to kilowatts simply isn't relevant to a demand that will be measured in megawatts to gigawatts.

Finally, the reality in North America, and the U.S. in particular, is that hydro dams are being intentionally breached (the reservoir walls are being broken) in recent years, reducing hydro generating capacity. Facing a rising demand for sustainable energy sources, hydro capacity is going in the wrong direction.

## 6.5. Algae

One of the potentially renewable biomass sources for hydrogen production that has received a high degree of media attention is the potential for huge ponds covering hundreds of hectares (or acres) to farm algae.

Algae is the plant form that grows in water, typically near the surface, and seasonally turns some waterways and reservoirs into a hideous bright green. These "algal blooms" are often an indication of excess phosphates in the water (pollution), and are encouraged by warmer water temperatures. The temperature factor has a strong hand in the seasonality factor of the blooms.

Scientists have discovered that under certain circumstances, certain species of algae have the ability to emit hydrogen. The theory is that by farming sufficient quantities of the best hydrogen-producing types of algae in shallow ponds, it will be feasible to harvest significant quantities of hydrogen. It has even been postulated that such operations could also clean up certain kinds of pollutants from the water in the process.

This theory sounds lovely at first blush, but as with so many of the hydrogen herd announcements, it glosses over certain inconvenient realities.

There is no consensus yet as to which varieties of algae would be best suited for the job. There is definite interest in crossbreeding existing algae to see if a suitable variety can be created. There is even the thought that if conventional breeding techniques are unsuccessful, genetic engineering can produce a super-algae for the job.

The algae does not produce hydrogen continuously. It only produces hydrogen under specific circumstances, for a relatively small portion of the overall time cycle. The algae will need to be managed intensively to create these conditions repeatedly and consistently. In another phase, algae actually will consume hydrogen, so the desired hydrogen will have to be harvested to ensure it is not accessible to the algae for consumption.

The infrastructure to grow the algae, collect the hydrogen, and then dispose of the dead algae will be huge and expensive. There is little work to date to indicate that such a technology will be cost-effective, even after we solve the technical issues to make it workable, let alone sustainable, on a commercial scale.

Even if all these obstacles are overcome, it appears likely that it would be more efficient and practical to farm algae to produce, not hydrogen, but biodiesel fuel.[7] [8] Biodiesel is another current darling of the renewables high-flyers.

Mike Briggs from the University of New Hampshire (UNH) Physics Department and UNH Biodiesel Group has been researching the topic of energy from algae, and currently focusing on research on algal biodiesel and other biofuels technologies. He writes:[9]

"The problem with producing hydrogen from algae is that the requirement of sulfur stressing the algae to get them to release hydrogen, then removing the stress to let them recuperate, makes it a slow process. On a per land area basis, you can produce more energy in the form of oil (or straight raw biomass) per year than you can in the form of gaseous hydrogen, because of the need to stress the algae (and they still only release a portion of their hydrogen), then remove the stress, and repeat. Just letting them grow, harvesting, and extracting the oil is simpler, more economical, and gives higher yields."

## 6.6. Biodiesel

Biodiesel ($C_{19}H_{36}O_2$) is not an attractive candidate for production of hydrogen. The hydrogen content ratio is quite low compared to other hydrogen carriers. It is still a relatively rare commodity. Its key advocates as a hydrogen source are presumably the agri-business sector looking for another market for its vegetable oil products.

## 6.7. Crop Wastes

There are basically three ways to use crop wastes to produce hydrogen. One is to burn the crop wastes to produce heat to drive a turbine to produce electricity to power electrolysis to produce hydrogen. It's really as cumbersome as it sounds. This process is effectively the same as for wood.

---

7.    *Widescale Biodiesel Production from Algae*, Briggs, Michael, University of New Hampshire, Physics Department
      http://www.unh.edu/p2/biodiesel/article_alge.html
8.    *A Look Back at the U.S. Department of Energy's Aquatic Species Program: Biodiesel from Algae*, U.S. National Renewable Energy Laboratory
      http://www.nrel.gov/docs/legosti/fy98/24190.pdf
9.    Personal correspondence.

The second option is to decompose the plant matter anaerobically to produce methane gas (biodigestion), and then reform (also referred to as "cracking") the methane to extract the hydrogen. The latter step is essentially the same process as is used to produce hydrogen from natural gas, which is primarily methane. This is not practiced on a significant scale today. What methane is produced by such processes today is typically used as methane fuel, and not as a feedstock for producing hydrogen.

The third option is to convert the crop wastes to ethanol via fermentation and distillation, or via cellulosic ethanol production. This is described in more detail below.

## 6.8. Ethanol

Production of ethanol ($C_2H_5OH$ or $C_2H_6O$) via enzymatic processes on cellulosic material shows a lot of promise as a cost-effective and sustainable means of making ethanol on a large scale. This process can use crop wastes as a feedstock, rather than more valuable food stocks typically used to distill ethanol.

However, in the event that large quantities of cheap, sustainable ethanol should become available, it is unlikely that reforming it to obtain hydrogen will be the most effective use of the ethanol from an energy efficiency or economy perspective. Instead, it will be better to use the ethanol directly in ethanol fuel cells currently being developed and scaled up to larger capacities, or as a fuel in internal combustion engines—either straight or blended with gasoline to extend existing supplies.

## 6.9. Wood

Wood is a high-carbon fuel source. It is labour intensive to harvest, cure, and handle prior to burning, especially on a large scale. So much so, there is very little large-scale power production from wood. Energy from wood is predominantly used for space heating.

## 6.10. Power Satellites

To my mind, this concept is the epitome of pie-in-the-sky wishful thinking. No question, there is an intuitive appeal in the idea of harvesting solar energy where the clouds and air pollution do not diminish the intensity of the sunlight. There are people sufficiently serious about the idea that the U.S. government has done significant research on the topic (e.g., NASA Space Solar Power Technology Advanced Research and Technology program) and held public hearings on the subject (House Science Committee Hearings on Solar Power Satellites).

In short, the technology isn't ready, and it is at least decades away even at the technical feasibility level. That's according to NASA, after five years of research on the subject.[10]

However, we must also consider the issues of energy or economic payback. Consider the energy consumed to produce the solar panels that will survive in the Earth's orbit, or to build the equipment required to convert that energy to a form that can be transmitted to the surface (e.g., microwaves), or the energy required to lift all that mass into orbit, and the earth stations required to re-convert the energy into something more usable. Further, the life of such orbiting energy stations can be cut short by malfunctions—servicing will not be a simple matter—or collision with the considerable debris we already have floating about in Earth's orbit. Most of that debris is too small to track by ground-based radar, but still very capable of doing significant damage on impact. Such satellites will presumably need to be in geo-synchronous orbit (in order to maintain a geographical link with its base station), a relatively small area that is already populated to some extent.

Other proposals suggest means of reducing the amount of mass to be lifted, by mining the materials on the moon, which presumes a base of operations on the moon that is not in anyone's short-term plans.

There are many other non-trivial issues. How do we identify the power beam corridors so that we don't have to worry about other flying objects intruding on them, to their own detriment? Do birds that fly into the beam simply get cooked, or do we implement some sort of intrusion detection that shuts down the beam when migrating butterflies or other insects enter the perimeter? While the power beams may be diffused to reduce the level of harm associated with incidental exposure, the exposure to a butterfly migrating through the zone (as one example) is likely to be appreciable.

Some of the attenuation factors in the atmosphere that reduce the amount of sunlight at the Earth's surface will also apply to solar power satellites in beaming their power to the surface, albeit to a lesser degree. There are also the conversion losses (e.g., PV to electricity, electricity to microwave, microwave to electricity).

10.  Statement of John C. Mankins, Manager, Advanced Concepts Studies, Office of Space Flight, before the Subcommittee on Space and Aeronautics Committee on Science U.S. House of Representatives, September 7, 2000
      http://www.spaceref.com/news/viewpr.html?pid=2568

Of the 1400 watts per square meter of solar energy available outside the Earth's atmosphere, there are serious questions as to what fraction of that would eventually be available to the end-consumer. Would it be appreciably more than we will get by installing PV panels on the roof of our buildings using the level of PV technology that will exist 50 to 100 years from now (an estimate of the earliest date when space-based solar energy could be available to us in any appreciable quantity)?

Surely, if this technology is to be made attractive in the future, the first step is to improve the efficiency and economy of photovoltaic technology in general, whether it is to be deployed on our buildings or in space. Until we address that issue, let's not start building the launch fleet.

## 6.11. Sustainable Sources in Aggregate

The amount of electricity we produce from sustainable sources on the planet today is small, and a shrinking fraction of the total worldwide electrical generation.

Outside of a few specific countries (notably in Europe), there is no viable government strategy in place to appreciably increase the amount of energy we produce from sustainable sources. In fact, the federal governments in Canada and the U.S. are actually reducing their funding for research, development, and implementation of true sustainable energy sources. Government subsidies to the fossil fuel and nuclear industries are faring much better.

Sustainable sources of energy are not going to be significant players in the production of the hydrogen required to sustain a planetary hydrogen economy. There isn't enough in place now, and there won't be for many decades to come, especially in the absence of a coherent and credible strategy to bring massive amounts of sustainable energy production to fruition in the near future. Instead, the future of hydrogen is clearly dependent on non-renewable energy sources, notably fossil fuels, for at least the foreseeable future. And while we know that is a dead-end road in the long-term, it does bear investigation for its short-term prospects.

# 7

# *Hydrogen From Fossil Fuel Sources*

While it is comforting to think that hydrogen could be produced from environmentally benign and sustainable energy sources, the present reality is very different. Today, most hydrogen is produced from fossil fuels, notably natural gas. As natural gas supplies diminish in North America, the heir apparent is not a sustainable energy source, but another fossil fuel—coal. This chapter reviews the primary fossil fuel candidates for hydrogen production, today and into the future.

## *7.1. Steam Reforming of Natural Gas*

Natural gas is effectively methane ($CH_4$), with a few impurities that are variable by the source, e.g., sulphur.

Methane is the primary constituent of landfill gas, swamp gas, marsh gas, and firedamp in coalmines.

Natural gas is the current predominant source for commercial production of hydrogen in quantity. Over 90% of the world's total production of hydrogen (as of 2003) comes from this finite, fossil fuel source.

The process of steam reforming natural gas to produce hydrogen ranges from 60% to 80% efficiency. Industrial scale installations tend toward 65% to 70% efficiency. While some industry sources claim efficiencies in the range of 85%, it appears they are considering only the amount of energy embodied in the hydrogen relative to the natural gas consumed directly in the process, not the energy required to drive the process itself. The key area of loss is the energy used (usually additional natural gas) to heat the steam reforming process to temperatures in the range of 1,000°F. Another facet of this technology that consumes energy is the pressurization of the process at anywhere from 3 to 25 bar (45 to 375 psi). High temperatures, high pressure, and flammable materials: do not try this at home!

An efficiency of 70% means the hydrogen produced contains only 70% of the energy embodied in the natural gas (and possibly other energy sources) to produce the hydrogen. In other words, we are discarding about 30% of the available high-quality energy in order to have the remainder in the form of hydrogen.

The process is typically performed in stages (multiple reactions), and aided by the use of catalysts, which are degraded over time, requiring replacement. The replacement process normally requires shutdown of the operation. Ignoring the multiple steps, the overall reaction is endothermic (absorbs heat energy) and can be described as:

$CH_4$ (natural gas) + $2(H_2O)$ (steam) => $4(H_2)$ + $CO_2$

Note the significant consumption of water (two water molecules for each natural gas molecule) and the production of the greenhouse gas, carbon dioxide $(CO_2)$, which are part of the reforming process.

Despite recent rises in the price of natural gas in North America, there is no worldwide shortage of natural gas. Natural gas could be diverted to hydrogen production in the future. It is still common practice in the oil industry to flare off natural gas at many oil wells (even in contravention of local laws). This is especially true of sour gas sites. Eventually, increased demand and higher prices for natural gas may make it financially worthwhile to capture these additional natural gas sources and process them, rather than just flaring them off into our air.

In the event that North American demand for natural gas should reach the point that it is considered too valuable to process into hydrogen, there are still larger reserves available elsewhere in the world. The existing oil companies will certainly be prepared to import natural gas to North America, just as they do with oil today, once the price is high enough to justify the investment.

An alternative would be to do the steam reforming closer to the natural gas source, and transport the hydrogen. Natural gas, however, is easier to transport than hydrogen.

Methane gas is another potential source of natural gas to be used as a production stock for hydrogen. It could be produced in large quantities from animal wastes, especially in the agribusiness sector. This would at least have the benefits of being a renewable resource, making better use of a waste product, and reducing effective greenhouse gas emissions (as methane is a more potent greenhouse gas than carbon dioxide) that would be produced from stripping the hydrogen out of the methane.

Some other hydrocarbons are examined below. An examination of all hydrocarbons is beyond the scope of this volume.

## 7.2. Methanol

Methanol ($CH_3OH$ or $CH_4O$) is also known as Bieleski's Solution, carbinol, colonial spirit, Columbian spirit(s), methyl alcohol, methyl hydrate, methyl hydroxide, methylol, monohydroxymethane, pyroligneous spirit, pyroxylic spirit, wood alcohol, wood naptha, or wood spirit.

While methanol can be produced from sustainable sources, its primary source today is fossil fuels.

Methanol is simply the alcohol version of methane, adding the oxygen atom to the methane molecule. Methanol has the advantage of being a liquid at temperatures that are friendly to people, while methane is a gas. Removing the hydrogen from a methanol molecule leaves a carbon atom and an oxygen atom. Left to their own devices these discarded items would form carbon monoxide molecules, which are poisonous to humans. Most likely, additional oxygen will be absorbed from the atmosphere to produce carbon dioxide, which is considerably less toxic, but a significant greenhouse gas.

Methanol is rare in nature, and generally has to be produced from other substances (methane, oxygen and hydrogen), using energy in the process. There are established processes for reforming methanol to produce hydrogen (enzymatic, catalytic, etc.). While research continues to try to find more efficient and cost-effective methods, it does not appear that anyone is yet using methanol as a hydrogen source on a commercial scale. Based on a reading of the literature, it appears the most likely source for methanol in the foreseeable future will be the use of coal.

## 7.3. Liquid Propane Gas

Liquid propane gas (LPG) ($C_3H_8$) is produced primarily as a by-product of natural gas production and the refining of oil products. In both cases, the propane is removed from the other materials as an impurity that could cause problems if not separated.

At room temperature, propane can be forced into a liquid state at pressures above approximately 7.5 atmospheres (7.6 bar or 110 psi).

While propane could be reformed into hydrogen by much the same process as natural gas, it is a less attractive candidate than natural gas for several reasons. Propane has a lower ratio of hydrogen to carbon than does natural gas (8:3 vs. 4:1). In other words, there is 50% more carbon to deal with in propane than in natural gas to arrive at the same amount of hydrogen. Propane is not as abundant as natural gas. Finally, the ability to store propane in compact, liquid form at relatively low pressures (those ubiquitous propane tanks ranging from sizes for propane torches to barbecues to household supplies) means there is a higher value market for propane than there is for natural gas. It takes more energy to liquefy natural gas, and liquefied natural gas (LNG) is stored at higher pressures than LPG, posing additional hazards.

So, while theoretically a candidate, propane is not as attractive as other hydrocarbons for use as a hydrogen feedstock. There does not appear to be much commercial interest today in using propane to produce hydrogen. There is, however, a 4-kilowatt demonstration fuel cell based on reforming propane in San Antonio, Texas.[1]

## 7.4. Coal

Coal is the main primary energy source for electricity generation in the United States today. It has earned its reputation for being an extremely dirty energy source. This reputation derives from several factors.

1. The mining of coal is a dirty business. It has led to the deaths of miners, not just from explosions, collapses, floods, and other accidents in the mines, but also from long-term health effects, notably "black lung".

2. The environmentally unfriendly techniques used, such as strip mining, decapitating mountains, and filling valleys with mine waste. This has led in many cases to the rapid erosion of the cleared areas and the silting up or blocking of waterways causing significant fish kills and impacts on other flora and fauna dependent on the affected watersheds.

3. The poor environmental record of the coal mining industry for returning closed mines to stable states, instead leaving hazardous substances to leach out over subsequent decades, poisoning groundwater and waterways.

Different grades of coal have different characteristics when it comes to burning them for fuel, which will determine how much energy can be extracted from a tonne of coal, or how much of various emissions will be generated from the

---

1.  Texas Commission on Environmental Quality press release
    http://www.tceq.state.tx.us/comm_exec/communication/media/1-04propanecell.html

burning of that tonne of coal. For example, the amount of sulphur in coal varies from source bed to source bed. This sulphur, when burned, is a major source of airborne sulphur dioxide, which has long been associated with acid rain and killing lakes in the U.S. northeast and eastern Canada.

Similarly, coal combustion is associated with the production of nitrogen-oxides and ozone, which are pre-cursors for smog. This production from coal-fired electrical generating stations in the U.S. mid-west, and central eastern seaboard, especially during warm summer weather is a key causal factor for smog-days in the U.S. northeast and eastern Canada.

If one is prepared to invest in various technologies to clean up the emissions (advanced burner technology, scrubbers, flue gas treatment, etc), it is possible to make coal burn very nearly as cleanly as natural gas generating stations do today.[2] This potential may become more important in years to come as natural gas becomes more expensive as demand outstrips supply while reserves of coal in North America remain relatively large. However, the coal burning generating stations actually in operation in North America today do not come close to these standards, and emissions standards in the U.S. have been relaxed from 2001-2004. This reversal is sufficiently dramatic that it has seriously affected the companies in the business of making emissions abatement equipment.[3] Of course, if North America increases its use of coal dramatically in the future, the reserves to production (R/P) ratio will fall as well.

As the North American coal-burning industry, notably electrical generation, has lobbied effectively to be exempted from meeting higher standards for clean air, they have become one of the major sources of air pollution on the continent. Just how major was brought home during the course of the August 2003 blackout in the U.S. northeast and Ontario. According to a study by the University of Maryland, just 24 hours into the blackout that took over 100 of the coal-fired plants in the U.S. off-line, associated emissions dropped dramatically. Airborne sulphur dioxide dropped by more than 90%; ozone by about 50%; particulates by about 70%; nitrous oxides by 80%.[4]

---

2.    Presentation by Keith Rivers at IEE/IEEE Electric Power Symposium IV, October 29, 2004, Ottawa

3.    *Relaxed Air Rules Choke an Industry*, August 31, 2003 Los Angeles Times article by Elizabeth Shogren, on Common Dreams Web site http://www.commondreams.org/headlines03/0831-03.htm

4.    *The 2003 North American Electrical Blackout: An Accidental Experiment in Atmospheric Chemistry*, Marufu, Taubman, Bloomer, Piety, Doddridge, Stehr and Dickerson, Geophysical Research Letters, May 2004

The industry's adamant refusal to embrace available technologies to improve its emissions and its illegal actions (e.g., under the Clean Air Act in the U.S.)[5] are the main reasons that clean coal is unlikely to be taken seriously going forward. The oft-stated position of the coal-burning electrical generation operators that it is technically difficult and too expensive to retrofit new, cleaner technologies to their plants is going to be the main reason that their industry will not be invited to be part of environmentally acceptable solutions for producing needed electricity in the future.

No matter how clean a coal-burning plant becomes regarding particulates, ozone, nitrogen-oxides, or sulphur-oxides, one undesirable by-product will remain a major challenge. As the world focuses more and more on the effects of global climate change, and the causality of greenhouse gases, the prodigious amounts of carbon dioxide produced from the combustion of coal will be a major concern to the world's inhabitants. I discuss more on carbon dioxide sequestration later in this chapter.

## 7.5. Clean Coal (FutureGen)

Clean coal has several incarnations. It is also referred to as coal gasification, and more recently it is the star of the FutureGen initiative of the U.S. federal government.

However, peering behind the PR façade, it appears the clean coal initiatives are more smoke than substance. Cleaner coal seems more appropriate, and given the current starting point, that does not present much of a challenge. Technology exists today that can do much to clean up the emissions (excluding carbon dioxide) from coal-fired operations. The catch is that it costs something to implement, either in new operations or as retrofits, and that cuts into profits. It is natural for plant owners to set out to meet only the minimum standards set by law, so long as that approach maximises their profit by minimizing their costs.

---

5.  In August 2003, FirstEnergy's Ohio Edison was found guilty of upgrading seven coal-fired plants and not installing required pollution control equipment. This followed similar judgements regarding dozens of other coal-fired plants throughout the U.S.

Let's have a look at the recent history of clean coal, starting with a major advocate, the United States Department of Energy (DOE), and specifically, their Clean Coal Technology Roadmap.[6]

1.   The U.S. DOE figures there is a 1000% return on investment to be had from the clean coal initiative over 15 years. Yet the coal and electrical generation industry isn't sufficiently interested in this financial miracle to go it on their own, they need 50% of the investment to be made by taxpayers before they're prepared to make this leap. Show me a mutual fund that reasonably expects that kind of return, and it can have my retirement savings tomorrow.

2.   Lots of talk about goals, but not too much about any real hard targets the industry is prepared to stand behind. Given the background and track record, it's hard to take this seriously. To create some credibility for this exercise, the industry should agree that, in return for the taxpayer stake, any plant that does not meet the efficiency and emissions targets set out for 2005, 2010, and 2020 should get shut down until it is enhanced to the point of meeting the targets.

3.   There seem to be two primary, but divergent, sets of goals in this roadmap. One is to reduce costs per kWh produced; the other is to clean up the emissions side. Based on the industry's history, where these two end up in conflict, the emissions targets will be found to be softer than the price targets.

Overall, I think US $10,700,000,000 would be better spent on conservation and demand-side efficiency efforts, and on sponsoring renewable energy projects than as seed money for projects to enhance cleaner coal combustion. Note, this money is not earmarked for actual implementation; just research, studies, development of technology and pilot projects.

Continuing with the U.S. DOE, consider the Techlines (New) link on their Web site (as of April 2004) accessed via their Fossil Fuels page:[7]

1.   Here we find more than 10 announcements about government funding for research and studies. There are none about implementations or

---

6.   The US DOE Clean Coal Technology Roadmap
      http://www.netl.doe.gov/coalpower/ccpi/pubs/CCT-Roadmap.pdf
7.   USDOE Fossil Fuels page Techlines (News)
      http://fossil.energy.gov/

results. Remember, concerns about cleaning up coal industry emissions date back decades, as do clean-up technologies.

2.  Shifting from government to the industry, there is the Coal Utilization Research Council, and its Web site, with its stated goals.[8] This page lists its accomplishments. It's all about getting taxpayer funds and tax incentives and getting favourable (less stringent emissions rules) legislation passed. Nothing concrete about reduced emissions, improved efficiency, or plants actually being upgraded or improvements to mining techniques or impacts.

Still in the industry sector, here are some gems from the Web site of the Center for Energy and Economic Development.

1.  "The U.S. Department of Energy estimates that utility industry compliance costs for reducing $SO_2$ and $NO_X$ have been reduced by approximately \$75 billion as a direct result of the success of the Clean Coal Technology Demonstration program".[9] That would be savings financed by taxpayers.

2.  Also according to the Center for Energy and Economic Development, $NO_X$ control technologies already developed and proven effective are not being implemented universally across the coal-fired generation industry.[10]

Shifting to Canada, there is the Natural Resources Canada (federal government) Clean Coal Roadmap Web page.[11] This Web site states that 17% of electrical generation in Canada comes from coal. Other sources indicate this figure is expected to fall in the near future (e.g., Ontario going to zero).

However, the most telling of these documents regarding clean coal may have to be the NRCan Clean Coal, $CO_2$ Capture and Storage Success Stories Web

---

8.  Coal Utilization Research Council
    http://www.coal.org/content/goals.htm
9.  Center for Energy and Economic Development
    http://www.ceednet.org/ceed/index.cfm?cid=7529,7591
10.  Center for Energy and Economic Development
    http://www.ceednet.org/ceed/index.cfm?cid=7529,7591,7685
11.  Natural Resources Canada
    http://www.nrcan.gc.ca/es/etb/cetc/combustion/cctrm/htmldocs/overview_e.html

page.[12] There aren't any success stories; just information about conferences, presentations, and future opportunities. That about sums it up.

Given the direct link between the FutureGen initiative and the U.S. administration's drive towards the hydrogen economy, I will deal with FutureGen in more depth. Compared to the overall spending forecast for clean coal at over US $10 billion, the spending for FutureGen is only a billion dollars. According to the U.S. DOE, Office of Fossil Energy (March, 2004):

"The production of hydrogen from FutureGen will support the President's Hydrogen Fuel Initiative to create a hydrogen economy for transportation".[13]

The U.S. government feels this initiative requires taxpayer funding because "the coal industry will not make the investments necessary to fully realize the potential energy security and economic benefits of this plentiful, domestic energy resource"[14], unless somebody else pays for the required upfront R&D. More likely, the industry has already done its own analysis and realized that given the efficiency penalties and costs of serious pollution abatement, coal simply isn't a viable energy source when environmental impacts have to be internalized.

"The FutureGen plant is planned to operate as a nominal 275 MW (net equivalent output) facility that produces both electricity and hydrogen and sequesters (captures) one million metric tons of carbon dioxide per year. Future-Gen will contain components which constitute high risk research activities" including "advanced generation coal gasification technology integrated with combined cycle electricity generation, hydrogen production, and capture and sequestration of carbon dioxide".[15] The project also includes a sequestration verification and monitoring component, although it is only scheduled to operate for three years (after eight years of work on the actual sequestration activity). Given the plan to sequester the carbon dioxide into geologic formations, three years of monitoring seems a trifle short compared to geologic time scales.

Very little of what FutureGen treats as pioneering research is actually new. Coal gasification dates back more than a century, and has been used in North

12.    Natural Resources Canada
       http://www.nrcan.gc.ca/es/etb/cetc/combustion/co2trm/htmldocs/
       success_stories_e.html
13.    *FutureGen; Integrated Hydrogen, Electric Power Production and Carbon Sequestration Research Initiative—Energy Independence through Carbon Sequestration and Hydrogen from Coal*, United States Department of Energy, Office of Fossil Energy, March 2004
14.    Ibid
15.    Ibid

America, Europe, and Africa in the past. Synthetic gas (or syn-gas) is hardly something new. The FutureGen desired product is analogous to the "city gas" of centuries past (a mix of methane, hydrogen, and various impurities). Power plants that can run on multiple fuels are also not a new technology—most boiler-based systems have this capability.

Underground sequestration of carbon dioxide is already a proven industrial practice in the oil industry (Sweden, Saskatchewan, etc.) to extend the life of oil fields.

The monitoring may be the most innovative aspect of FutureGen, but it appears to be an afterthought for the project. As for the degree of innovation involved in monitoring, consider that commercial $CO_2$ detectors for industrial use have been in use for years.

According to the estimates of the sponsoring agency, it will be at least 2018 before we will know if the project is viable, and presumably well into the 2020s and 2030s before it becomes a commercial reality, if it is proven successful.

In short, the FutureGen initiative seems little more than a coal-black hole to consume another billion dollars of U.S. taxpayers' money in order to generate more hydrogen economy hype. It also appears to be a PR vehicle for the coal industry, permitting it to further delay making real advances in cleaning up their operations in the short term. One thing it won't do is provide useful solutions within the next human generation, and probably two.

As with many of the proposed hydrogen production technologies based on fossil fuels, clean coal needs to address the issue of carbon sequestration.

## 7.6. Carbon Dioxide Sequestration

Due to increasing concerns about the impacts of greenhouse gases in our atmosphere, there are significant concerns about the amount of carbon dioxide that would be emitted from the production of hydrogen from hydrocarbons, and subsequently released into the atmosphere. For example, when natural gas is steam reformed to produce hydrogen, the primary by-products are heat and carbon dioxide. When confronted with this issue, proponents of the hydrogen economy typically respond with the carbon dioxide sequestration solution. So, what is it?

Well, in simple terms, it just means burying the carbon dioxide. Unfortunately, as carbon dioxide is a gas, burying it is not a simple thing to do successfully and for extended periods of time. There is no lack of folks willing to do more paid research on the subject, which is interesting given that this is an established industrial practice in the oil industry today. At several oil wells, oil companies are already pumping carbon dioxide gas into partially depleted oil fields to increase the pressure in the underground deposits, so as to increase the amount of oil that they can economically retrieve from the site. Such operations have been carried out by Statoil (Sweden) and in Canada at wells near Weyburn, Saskatchewan. Clearly, as this is already industrial practice, we do not need to do additional research on how to pump carbon dioxide into the ground. We do need credible research on how effectively the carbon dioxide stays buried. While there is some recent interest in researching this field, it will take decades or more before we know if this can be done effectively and safely in the long-term (for several human generations). Given the practice is five years old, it seems a little premature to declare these two demonstration projects successes. After all, the Weyburn monitoring project is expected to take at least 30 years, and the objective is to store the carbon dioxide in a stable manner, with no leaks, for geologic periods of time (hundreds of thousands of years).

The catch is that carbon dioxide doesn't like to stay buried. It is lighter than the oil in the fields, so if there is a hole in the ground, the carbon dioxide is likely to find it and escape into the atmosphere. So, while the oil industry has been pumping carbon dioxide into the ground for years, only recently has it begun to study how well it actually stays there.

Andrew Renton has written papers on the hydrogen economy and carbon sequestration. He has written:[16]

"If we generate our hydrogen by reformation from a fossil fuel such as coal, we have to sequester, otherwise we make our greenhouse gas emissions even worse. If we are sequestering carbon dioxide waste into geological formations, who writes the contract that says when the statute of limitations is up on it being contained? Is 100 or 1000 years alright before it leaks? What about 10 or 20 years? Who decides what's an acceptable period of time?"

"This is just sealing the nasties up in a plastic bag for a finite period of time. At some point it will leach out."

Sequestration of carbon dioxide gas is not a permanent solution. The gas will eventually leak out, and be just as potent a greenhouse gas then as it is now. However, there is a potentially even more dangerous consequence from carbon dioxide sequestration: sudden release of the sequestered gas.

Let us suppose the gas is sequestered underground in a stable geological formation. Then, an unforeseen geophysical event, such as an earthquake, volcanic

---

16.  New Zealand Energy Efficiency and Conservation Authority, Andrew Renton
     http://www.eeca.govt.nz/content/EW_NEWS/82june03/plentiful.html

eruption, tectonic shift, or badly placed drill-hole provides an outlet to the gas so that it is released quickly (in days rather than decades) and in a large quantity. Carbon dioxide is toxic to humans in high concentrations. What happens in this event?

In fact, this very experiment has already been conducted, albeit accidentally and with terrible results. In 1986, at Lake Nyos in Cameroon, a large carbon dioxide bubble vented to the surface. While lighter than water and the planet's crust, carbon dioxide is not lighter than air, and it hovers at ground level until it disperses. At Lake Nyos on August 26, 1986, these properties resulted in the asphyxiation of approximately 1,700 people. No survivors. A similar event at Lake Monoun in 1984 killed 37.

While the mass killings at Lake Nyos and Lake Monoun resulted from natural phenomena, what needs to concern us in this discussion is that large releases of carbon dioxide gas have the ability to kill large numbers of people. Because the gas is odourless and colourless, it will likely arrive without warning. It won't matter if the release comes from a man-made carbon dioxide sequestration project or one performed by Mother Nature. When it happens, people in the area will die. Even if monitored, how much warning will be provided? What measures will be in place?

Hopefully, the monitoring will include extensive seismic monitoring, including baseline surveys, before sequestration operations begin. There is substantial evidence that show the pumping of wastewater in quantities smaller than envisaged for carbon dioxide sequestration has caused earthquakes at the Rocky Mountain Arsenal near Denver in the 1960s and at Ashtabula Township in Ohio in the late 1980s and into the 1990s.[17] This leads me to wonder if the actual act of pumping large quantities of carbon dioxide into geological formations could cause the very seismic disturbance that could in turn lead to the release of the sequestered carbon dioxide to the atmosphere.

If sequestration of carbon is to be part of the technological solution to hydrogen production from hydrocarbons on a large scale, sequestration of carbon dioxide gas in geologic formations should not be the solution employed.

I won't be moving my family close to areas where carbon dioxide is being pumped into the ground. Certainly not near a FutureGen site that plans to

---

17.  Geotimes
     http://www.geotimes.org/mar02/NN_quakes.html
     For more citations on the seismic effects of deep well pumping, check the Induced Earthquake Bibliography at
     http://www.nyx.net/~dcypser/induceq/induceq.html

sequester a million tonnes of carbon dioxide per year, with no long-term monitoring plan.

There are also proponents of other means of sequestering carbon dioxide. I don't have space here to catalogue all the means suggested to date, but will cover a couple that are getting some attention.

Growing more vegetation has received some play, especially in environmental advocacy and agricultural quarters. The basic idea is to create more vegetable matter (e.g., trees) that capture carbon dioxide from the atmosphere as they grow, thus removing it as a greenhouse gas. That seems intuitively reasonable, but it works only so long as the plant material remains in a form that continues to hold the carbon, e.g., as furniture or housing stock. However, as soon as the material is burned for fuel or composted or processed by any other means that releases the carbon back into the atmosphere to become carbon dioxide again, the benefit is lost. Trees have the advantage that they sequester more carbon per square foot of land surface due to their height, but suffer the disadvantage of taking up to a century or more to reach maturity. Another significant downside is that a forest fire can wipe out the benefit of years of sequestration in a matter of days.

A variant of the increased vegetation idea is ocean seeding, or subduction zone sequestration. Again, there is an intuitive appeal to the idea of enriching the environment for the growth of ocean-based microorganisms that will consume carbon, and then sink to the bottom of the ocean, in effect taking greenhouse gases with them. On a precautionary note, one would hope we would give some serious thought to how we have abused the oceans already, and the known effects from using them as a giant garbage pit. We must be very sure of what we are doing before messing with them in this way. However, that does not appear to be the case. Those that have given the matter serious consideration have concluded that we don't have a good grasp of the science, and we are not ready to start experimenting.[18] We don't really understand how subduction zones work, but we know they are a key factor in earthquakes. It would seem prudent not to mess with these zones until we have a much better comprehension of them.

A far better solution would be the sequestration of simple carbon, also known as carbon black. Simple carbon is a solid, and one that is used commercially. Carbon black can be produced from carbon dioxide by breaking the chemical bonds (which takes some energy), producing the carbon black, and oxygen. In fact, in

---

18.   *Is Ocean Fertilization a Good Carbon Sequestration Option?*, Adhiya, Jagat and Chisholm, Sallie W., 2001
      http://lfee.mit.edu/publications/PDF/LFEE_2001-001_RP.pdf

reasonable quantities, it would not be necessary to sequester carbon black in huge underground deposits. We could simply sell it on the open market for regular commercial use. Additional demand for carbon black could be generated by increased use of carbon fibre materials. Such materials are light, strong, and durable. One potentially large market for such material would be the automotive sector, as carbon fibre components could replace heavier metal parts in many applications (chassis, suspension parts, body panels, component housings). This would result in much lighter vehicles that are at least as strong and durable as those built today. It would also result in reduced greenhouse gases (GHGs) from a reduction in demand for steel and in reduced GHG emissions resulting from reduced (fossil) fuel consumption. The automotive sector is already a large consumer of carbon black—it is used in large quantity in tires. Much of the carbon black produced today comes from the incomplete combustion of natural gas. As the price of natural gas continues to rise, perhaps industry will eventually see the splitting of carbon dioxide into oxygen and carbon black as a viable alternative.

However, it takes energy to separate the carbon from the oxygen in carbon dioxide, and to date this option is more expensive than simply burying the problem, and therefore not as interesting to industrial concerns.

Nature's solution to sequestering carbon so the planet could provide an environment hospitable to us was to turn it into oil, natural gas, peat and coal. Leaky pressure vessels are not nearly as good a solution as binding the material at the molecular level.

## 7.7. Fossil Fuels in Aggregate

Fossil fuels are the predominant source for hydrogen today (steam reforming from natural gas), and the plan for the future (FutureGen and clean coal). Given the trivial amounts of sustainable energy available now and decades to come, and the dependence of current and planned hydrogen production on fossil fuels, the hydrogen economy as currently planned will rely almost entirely on dirty fossil fuels, predominantly coal.

Aside from the environmental issues that inevitably arise from our continued and increasing use of fossil hydrocarbons, there is one other reason to believe we won't be able to base a pervasive, long-term hydrogen economy on fossil fuels. They're running out. Fifty years from now, when the hydrogen economy might finally be blossoming (if all the research and development wishes come true), oil, natural gas, and even coal will simply be considered too precious to spend on making hydrogen.

# 8

# *Hydrogen From Other Energy Sources*

So far, we have established that current and foreseeable sustainable energy sources are insufficient to meet present or future needs, with or without the hydrogen economy. We also know that fossil fuels are not sustainable, probably not even until the hydrogen economy can be ready. As a result, many people have decided we need to explore other options for producing hydrogen. Let's examine some of those.

## 8.1. Nuclear Fission

Promoters of the hydricity concept, such as Geoffrey Ballard[1] and David Scott[2], expect that hydrogen will be produced in massive quantities primarily via electrolysis using electricity produced by nuclear fission reactors. I have covered the electrolysis component previously, but would like to focus further on the nuclear fission aspect.

The hydricity proponents make statements along the lines "hydrogen and electricity will become so indistinguishable from each other in the future that they will be referred to as a joint currency called hydricity."[3] However, it is a poor currency indeed that exacts up to a 50% transaction fee each time you wish to change modes.

---

1.  *Hydricity, the Universal Currency*, Geoffrey Ballard in Fueling The Future
2.  *The best defence against terrorism*, Scott, David Sanborn, October 6, 2001 Globe and Mail article, at University of Victoria Web site
    http://www.iesvic.uvic.ca/comments/
    The%20best%20defence%20against%20terrorism.pdf
3.  *Hydricity, the Universal Currency*, Geoffrey Ballard in Fueling The Future

Nuclear fission technology fell from favour in the 1970s, primarily due to the incident at Three Mile Island and the resonance of the movie, the China Syndrome, released at about the same time. Many believed the partial meltdown and radiation escape at Chernobyl drove the nails into the coffin of nuclear powered electrical generation.

However, the nuclear fission power production industry has not died. It has continued to produce reactors for the military (e.g., power plants for submarines and large warships). It has continued to build power reactors for other countries. It has effectively built new reactors in North America by means of upgrading many of the existing reactors already in place prior to 1973. Since 1999, the industry has been selling an average of three new reactors per year in North America.

Today, this industry appears poised for resurgence in North America. As concerns regarding climate change cause us to cast about looking for energy sources that emit fewer greenhouse gases, the nuclear fission industry trumpets that they are a green energy source with no greenhouse gas emissions.[4][5][6] These, and other proponents of the hydrogen economy point to nuclear energy as the preferred energy source for producing the hydrogen.

In fact, nuclear fission is not a zero GHG emissions energy source, when you consider the effects of building the facilities, mining and milling the uranium, building the fuel packages, and, energy consumed in operation and decommissioning of these plants.

Traditional nuclear fission generating plants (heavy water, pressurized light water) are huge consumers of concrete and steel, both of which are prodigious contributors to GHG emissions.

Uranium is not a renewable resource. Supplies are finite, and the supply of rich ore is being seriously depleted already.

Those who claim that nuclear fission is renewable are usually using the term to disguise their support for breeder reactors. Breeder reactors are designed to run on increasingly hazardous nuclear fuels (e.g., plutonium), which are by-products of the conventional fission process. This does not make the process renewable. It

---

4.    Canadian Nuclear Association
      http://www.cna.ca/english/climate.asp
5.    Nuclear Energy Institute
      http://www.nei.org/index.asp?catnum=2?catid=265
6.    Foratom Web site, "Nuclear Power and Climate Change"
      http://www.foratom.org/content/Default.asp?PageID=752

merely extends the life of the cycle, at the expense of producing even more dangerous by-products.

Nuclear fission is not clean. The mining and refining of uranium and processing of nuclear fuel is a dirty, hazardous business.[7] Spent nuclear fuel is not benign. Isn't it ironic how industrialized society can talk about nuclear waste as the primary constituent of "dirty" bombs, while also describing the process that produces the waste as "clean"? Sadly, our societies also accept distributing depleted uranium waste via spent munitions as a viable disposal option for deadly, spent nuclear fuel. Would the leaders get it if the nuclear waste repository were being proposed for Kennebunkport or Crawford, instead of Yucca Mountain?

There is also a question of whether or not nuclear fission is a net energy producer, or a net energy sink. It has been suggested that the answer to this question is largely a matter of the richness of the raw ores mined to provide the uranium. As suitably rich ores play out (and they are not plentiful even now), it will actually take more energy to build, fuel, and decommission a nuclear fission generating plant than it will produce.[8]

Still others have concluded that as the overall quality of uranium ores diminish over time (higher quality ores are exploited, leaving lower quality ores to be mined), nuclear fission may not even manage to reduce overall carbon dioxide (greenhouse gases) emissions.[9]

Ignoring the question of rich vs. poor uranium ores, there are others who make the case that there is not enough uranium on the planet to fuel new nuclear fission reactors on the scale required to meet projected power demand for the remainder of this century. Even the nuclear industry indicates that estimated reserves (excluding uranium dissolved in seawater that will be very expensive to extract) are good only for a few hundred years, at current rates of use.[10] If utilization rates increase dramatically, then the number of years those reserves will last must drop accordingly. The Nuclear Energy Association (NEA) scenarios end at 2050, and provide no assurance that there is enough uranium available to meet forecast needs in a growth scenario (without resorting to other nuclear fuels, e.g.,

7.    *Dying for a Living*, Tataryn, Lloyd
8.    *Is Nuclear Energy The Answer?*, Allison Macfarlane in <u>Fueling the Future</u>
9.    *Nuclear Power: the Energy Balance*, van Leeuwen, Jan Willen Storm and Smith, Philip
       http://www.oprit.run.nl/deneen/
10.   Nuclear Energy Agency
       http://www.nea.fr/html/ndd/climate/climate.pdf

thorium, plutonium). In fact, uranium looks like a sustainable energy source only by comparison to fossil fuels, which we know we are depleting rapidly.

This observation raises the spectre of breeder reactors using fuels that are even more radioactive, virtually all of which are nuclear weapons grade materials.

The nuclear industry's simplistic view conveniently ignores the track record of the nuclear power industry to date, and its on-going potential for catastrophic consequences.

Most of the nuclear reactors in North America are more than 25 years old. The average age is approaching 40 years, the originally anticipated effective life for these irradiated behemoths. There is still no effective strategy in place for the true decommissioning of reactors that have exceeded their rated life. It appears many of these reactors will be permitted to continue operation beyond the rated life of their components, increasing the probability of a serious incident, simply because there isn't a plan for how to close and dismantle them. Further, thousands of tons of highly radioactive spent fuel remain in cooling pools at reactor sites around the world because there is nowhere else for this waste to go (except for another reactor site and its spent fuel pool).

These aging reactors with their co-located spent fuel pools are monuments to our colossal technological arrogance, and willingness to reap short-term benefits while leaving the responsibility of dealing with the consequences to following generations. Could we ask for a better example of what does not constitute a sustainable solution?

Every country that has developed nuclear weapons capabilities has done so based on extracting the nuclear materials required from the spent fuel from their domestic nuclear power generation or research reactors. The number of countries that have a nuclear power generation reactor, and have not proceeded to demonstrate nuclear weapons capability within a decade is very small. This is not to say they don't have nuclear weapons, only that they have not publicly demonstrated them, or, are in a position to provide weapons material to other states.

One state that has a significant nuclear energy program (52 reactors in operation in 2004) and has not announced having nuclear weapons is Japan. However, some Japanese politicians have openly disclosed that it would not be difficult for Japan to produce nuclear weapons from their spent fuel. Japan was also a big promoter of fast breeder reactor technology until the accident at Monju at 1995. This was followed by issues arising from the proposed use of MOX fuel, which proved to be of sub-standard quality, and therefore too dangerous to use. There were also fears about the aging fleet of existing reactors, which was accentuated by the fatal accidents at the Tokaimura nuclear fuel conversion plant and the

Mihama No. 3 nuclear generating station in 2004. Even with this record, Japan is considered a leader in safety in a nuclear power program.

Spent fuel does not have to be fashioned into a high-tech fissionable device to be effective as a radiological weapon. It is far simpler, and nearly as effective, to gather a quantity of the radioactive material and disperse it with a conventional explosive in a populated area to inflict massive fatalities and illness from radiation sickness.

There are already tonnes of weapons grade nuclear materials in storage on the planet, and enough of that is unaccounted for to be a matter of concern to experts.[11] Why would we want to be producing more of these toxic, radioactive, carcinogenic materials even if they weren't nuclear weapons feedstock? The use of depleted uranium in conventional weapons is going to leave battlefields poisoned for decades, and more likely centuries, to come. The health effects tolls are only beginning to be understood from the depleted uranium rounds expended in the first Gulf War.

In the early years of promoting nuclear energy, the advocates were known to say that they would produce electricity at such low cost, it would be too cheap to meter (i.e., it would be free to consumers). In reality, nuclear energy is destined to be the most expensive electricity ever produced by the utility companies. When the utilities received a subsidy under electricity deregulation rules regarding "stranded costs", these were primarily about the costs they had sunk into nuclear generating plants, and would never be able to recover in a competitive marketplace against less expensive power production options. The reactors did not reach their predicted life spans without massive, expensive upgrades and refits.

Nuclear energy remains heavily subsidized by taxpayers.[12] We have not seen the final tally yet, as we do not know what the price tag for decommissioning the existing reactors will be, or the cost for disposal of the radioactive waste. However, initial estimates indicate that these costs will be a significant drain on our future economy.

Perhaps the most damning issue for nuclear fission is that we have no acceptable solutions for the permanent disposal of spent fuel and related radioactive wastes, despite the fact we have been producing it for over 50 years.

---

11.   *Database exposes threat from 'lost' nuclear material*, Trei, Lisa, Stanford Report, March 6, 2002
      http://news-serve.stanford.edu/news/2002/march6/database-a.html
12.   Canadian Coalition for Nuclear Responsibility
      http://www.ccnr.org/subsidies.html

There is only one location in the world today that is purportedly a permanent disposal site for spent nuclear fuel. It's called Chernobyl.

And the sarcophagus at Chernobyl is reported to be failing.

In spite of all this, construction licences for new nuclear fission plants are being issued in North America again. Beyond this resurgence of interest in conventional light water reactors (LWR), many hopes are being pinned on the potential of pebble bed modular reactors (PBMR). To read the promotional literature, PBMRs are the perfect energy solution, and address all the past issues with nuclear fission. Further, the promoters would have us believe that the research and development to date of PBMRs has been an unblemished record of successes. Well, you know the old saying, "if it sounds too good to be true, it probably is." It applies here.

First, PBMRs don't address significant issues related to existing fission technology. For example, the use of enriched uranium (weapons grade) as the fuel for PBMRs, or the matter of spent fuel remain unresolved. PBMRs will still produce it, and the industry (let alone governments) still doesn't have a viable long-term solution for storage or disposal of spent nuclear fuel. But PBMR proponents have gone further.

They have claimed the technology is so safe that (expensive) *containment buildings aren't necessary*. Given that PBMRs are based on using high temperatures and high pressures, and lighter than air gas (helium) as their coolant and moderator, that's an intriguing stance to say the least. Further, given that PBMRs are based on using enriched uranium encased in a shell of graphite, which is flammable, the potential for a release of radioactive material in the event of an accident seems more than a remote possibility.

Finally, the PBMR really does not have a reassuring track record.[13][14] You see, just because you change the acronym doesn't make the reality go away. The PBMR is just the latest moniker for the same basic technology that has previously had an alphabet soup of ancestors, e.g., advanced gas-cooled reactor (AGR), high temperature reactor (HTR), high temperature thorium reactor (HTTR), high temperature gas reactor (HTGR), thorium high temperature reactor (THTR), modular high temperature gas reactor (MHTGR), and gas turbine modular high

---

13.   A good review of the various reactor types leading up to the PBMR by Arjun Makhijani can be found at the IEER site, specifically at:
      http://ww.ieer.org/comments/energy/chny-pbr.html

14.   *Arguments on the Construction of PBMR Reactors in South Africa*, Steve Thomas, University of Sussex
      http://www.ratical.org/radiation/PBMR.html

temperature reactor (GT-MHTR). To learn the true pedigree of this technology, you have to trace back through the lineage.

In Britain, the AGR has never been permitted to refuel in operation (a purported advantage of the PBMR and its relatives), due to safety concerns.

The Dragon HTR (in Britain) operated for about 10 years, and was known to leak radiation. The experience was sufficiently bad that Britain stopped work on HTRs after the Dragon.

The first HTR reactor in the U.S. was Peach Bottom 1. It operated for only seven years, and was decommissioned.

The Fort St. Vrain HTR reactor in the U.S. was in commercial operation for only 10 years. It had an appalling record of equipment problems, and its average load factor over the ten years was just 15%. 80% is typical for conventional light water reactors.

The experimental HTTR plant at Oarai (Japan) has not lived up to expectations, and after significant accidents at the Monju reactor in 1995, and the fuel processing plant at Tokaimura, government interest has cooled considerably.

The THTR-300 in Germany is possibly the closest unit to the current PBMR design to actually become operational. HTTR advocates claimed the THTR design was inherently safe. So it is instructive that this unit was also plagued by equipment problems. Advocates have also claimed that the THTR-300 was shut down due to the accidents at Three Mile Island (1979) and Chernobyl (1986). An accident at THTR-300 on May 4, 1986 resulted in the release of radiation. The accident was the direct result of one of the fuel pebbles becoming lodged in the pipe feeding the fuel into the reactor. Circulating fuel pebbles are the key feature of the PBMR design. Officials attempted to cover-up the accident by blaming the fallout from THTR-300 on Chernobyl, which had its major accident the previous month. Despite this, the THTR-300 went into commercial operation in 1987 (after the Chernobyl accident) and remained in service until 1989. However, it appears that it never operated over two-thirds of rated power, and was out of service for months at a time, on more than one occasion, during the less than two years it was licensed for commercial operation. It has since been decommissioned. Its demise had nothing to do with Chernobyl.

Clearly, the predecessors to the PBMR technology have a significant and troubled history.

Interestingly enough, the PBMR advocates have been crowing that the high temperature aspect of these reactors will make them especially useful for producing hydrogen, by virtue of making it easy to break the water molecules into hydrogen and oxygen.

As the world's fleet of fission reactors is aging, more and more stories are coming to light of close-calls. As was reported in Wired magazine[15], nuclear plants are not aging gracefully. Perhaps it is simply time to accept that humans are not up to the challenge of using the nuclear fission cycle (from mining to spent fuel disposal) safely. We only have to get it seriously wrong once.

## 8.2. Nuclear Fusion

Given the amount of money that continues to be spent on research, you may be surprised to learn that nuclear fusion already works. It is a proven and robust energy source, and a significant fusion reactor has been in operation for some time now.

You've probably seen it; we call it the Sun. However, this is a little large for stuffing into a research lab or generating station. Corporations are still wrestling with an ironclad means to charge the general consumer for the energy provided from the Sun. So, the hunt continues for some way of producing a miniature equivalent of a star that we can harness and control, and if possible, actually produces more useful energy than it consumes.

However, the current reality is that nuclear fusion on a sub-planetary scale doesn't work now after decades of research and work, and it is not expected to work anytime soon. Still, expensive research continues.

While excitement regarding cold fusion waxes and wanes over the decades, the current reality is that it has not been produced in a repeatable manner under laboratory conditions, let alone producing energy in any useful quantity. There is no reason to believe it will be a viable energy source within the next century.

## 8.3. Methanol Plant

I am aware of a facility in British Columbia that produces methanol as its primary product. It also produces some hydrogen as a by-product. This hydrogen is used to produce ammonia, not to provide power to the plant. The current reality for hydrogen is that it is more valuable as a feedstock to produce other products than as a fuel itself.

---

15. *Nuke Plants Aging Disgracefully*, Kendra Mayfield
    Wired Magazine, February 2003
    http://www.wired.com/news/technology/0,1282,57486,00.html

These plants need energy to operate. Why aren't they capturing the hydrogen they are producing as a by-product, putting it into a fuel cell, and using the electricity themselves? Only one reason occurs to me—it isn't as economical for them to make this investment as to simply buy their power from their local utility or produce the electricity themselves from other sources. If companies that can get hydrogen for free are not prepared to make the investment to use it as fuel, how can we imagine it is going to make economic sense for the rest of us to buy hydrogen to use as a fuel?

## 8.4. Other Potential Hydrogen Sources

There are other means of producing hydrogen, other than via electrolysis or from renewable sources (covered later in this chapter). The lure of a piece of the tax money being thrown at the hydrogen economy is succeeding in finding more methods, including:

- Mineral acids on metals
- Thermal decomposition of aluminum hydride
- Steam with hot coke
- RECO-solar
- Treatment of wastewater
- Sodium hydride
- Bio-inspired catalysts (to split water)
- Nanotechnology (photon absorption with electrochemical splitting of water molecules)

I am not covering these methods (or undoubtedly dozens of other chemistry curiosities I have not mentioned) in the interests of space and readability. None of them are used on a commercial scale, and are not expected to be in the future. They are simply symptomatic of the noise level being produced by all the hydrogen hype. The feeding frenzy for government grants ensures that more announcements will be forthcoming, each claiming to be *the* answer. Or, at least it will be after a few more years or decades of funding for research and development. Spending time on each and every candidate just deflects us from the more important task; finding a better solution than the hydrogen economy.

It does not fall to me to debunk each wild-eyed advocate's solution to some piece of the hydrogen economy jigsaw puzzle. The market for hydrogen exists today for multiple industrial applications. If you have a process that can produce hydrogen cheaply and sustainably, what's stopping you from getting rich by commercializing your process today? Nothing. You don't require grants from the taxpayers to support your activities. The capitalist system exists precisely to finance this sort of advance.

However, now that we have examined the various means that have been postulated for producing hydrogen, from sources conventional to far-fetched, and both sustainable and not, the conclusion to the exercise is clear. There is no magic source for cheap, sustainable hydrogen on a massive scale on the horizon. As much as we might wish for such a thing, the reality is that we are stuck with what we know.

## 8.5. The Most Abundant Element

It doesn't matter how much hydrogen there is, because it is not an energy source. The fact that hydrogen is the most abundant element in the universe, as well as here on planet Earth is irrelevant when it comes to the topic of practical energy production, storage, and use.

Hydrogen is not found in a free state on this planet. It actually takes energy from real sources to produce hydrogen so that it can be used as a fuel.

# 9

## *Using Hydrogen*

Hydrogen is not a newly discovered material, nor are its storage and fuel cells. In fact, the hydrogen fuel cell is older than the internal combustion engine or the electric motor. See Appendix A, titled Hydrogen Timeline, for more on the history of hydrogen.

The degree of hype surrounding potential new means for producing hydrogen might make one think it is a rare commodity with limited uses. In fact, hydrogen is in widespread use today in a number of industrial sectors, and is already a mature industry on a significant scale.

One major use of hydrogen today is in processed foods. Check the labels on foods that you buy for the words "hydrogenated oils", "partially hydrogenated oils", "trans fats", "trans fatty acids", or similar. These are all indications that hydrogen has been used to process the food. This is almost always associated with industrial-scale food processing. The process is normally used to convert polyunsaturated liquid oils into solid fats, to reduce rancidity and make them easier to handle. Hydrogenated oils were frequently substituted for tropical oils (e.g., palm oil, coconut oil) when tropical oils were labelled as dietary villains a couple of decades ago. Hydrogenated oils are common in some margarines, vegetable oils, and many processed foods.

Such labelling and processing should become less common in the future, as the medical industry and health advocates have pressured many governments into forcing accurate labelling of these ingredients, and publicity around the fact that these hydrogenated fats in our diets are hazardous to our health. Some jurisdictions (e.g., Denmark, Canada) are actually banning the use of trans fats for health reasons. They are a prime cause of heart disease, indicted as leading to heart attacks, arteriosclerosis, strokes, etc. Curiously, one type of vegetable oil that is recommended for healthy diets as a good fat, canola oil, with its omega-3 component will also come up again as a feedstock for biofuels.

Hydrogen is used in the refining and production of petrochemical products. In fact, it is used to upgrade lower quality (heavier) feedstock to make them of higher value, putting the "hydro" in "hydrocarbon", so to speak. Given that we are exhausting the sweet (i.e., cheap), conventional oil sources around the planet as you are reading this, we may wish to conserve our hydrogen production capacity for upgrading the heavy stocks that will increasingly become part of our base supply of oil.

Hydrogen is used in the production of ammonia and ammonia-based fertilizers. It is used in food processing (to make margarine and a wide variety of processed foods as noted previously). It is used to produce polypropylene and other consumer materials. It is used in the manufacture of electronics products. It is used in scientific laboratories. It is used as a coolant in some extreme environments. It is used in the manufacture of some metals and fabrication processes. It is used to fuel rockets and the space shuttle. Although this is not an exhaustive list, it presents some of the major and high profile uses for hydrogen today. It is not a new commodity, nor a fledgling industry.

Despite these multiple uses of hydrogen today, the anticipated hydrogen economy will dwarf this current level of utilization to the trivial. Who is contemplating the impacts on the existing hydrogen-consumers when demand for hydrogen as a fuel competes with these applications? Will they disappear, much as the North American fertilizer production industry has subsided as the price of their primary feedstock, natural gas, has become too expensive for them to remain competitive?

Still, the existing hydrogen industry is mature, robust, and stable. It is reasonable to presume that the practices that are widespread in this sector are the result of extensive commercial and industrial experience, and the desire of the companies in that industry to be efficient and cost-effective (profitable).

What is particularly interesting is where the industry's current practices differ from the vision of the hydrogen economy advocates.

Where the hydrogen herd seems determined that hydrogen will be delivered by pipeline, and stored as a high pressure gas, the existing industry prefers to liquefy hydrogen for transport and storage, and avoids high pressure storage. Final delivery to customers in small quantities is typically in heavy tanks, storing the hydrogen at moderate pressure and ambient temperatures.

Where the hydrogen herd illustrates a future in which hydrogen is produced from clean, renewable sources, the industry today produces hydrogen almost entirely from non-renewable, polluting, fossil hydrocarbons.

Where the hydrogen herd postulates that hydrogen will be an economical fuel, currently it costs from four to eight times (depending on assumptions) as much as gasoline for the equivalent amount of energy (from steam reforming of natural gas). If the hydrogen is to be produced via electrolysis, it is likely to cost at least double that again. This is when demand is relatively low, and natural gas feedstock is relatively cheap. Advances in technology may bring prices down somewhat. However, it seems unlikely that a process that requires so much primary energy to extract a fuel will ever be competitive with current fossil fuels that are essentially extracted from the ground in near ready-for-use form.

In my view, there are two key aspects to the use of hydrogen: distribution and storage to make it available where it is needed; and, the actual consumption of the hydrogen to do useful work (including the devices used to perform the conversion).

## 9.1. Distribution and Storage

Hydrogen's physical characteristics make it a challenge to store. Not only is it a gas at all natural temperatures on earth, it is the universe's smallest atom and molecule. It simply leaks out of containers considered non-porous for most other substances. When freed into the atmosphere, hydrogen either combines with other substances or floats away. It does not puddle or pool. It is one of the most flammable substances known; it can ignite at concentrations ranging from just 4% up to 75% if even small quantities of oxygen are also present. It reacts with many metals, which can lead to embrittlement. As a gas, the element has a very low density at atmospheric pressure, leading to a low energy density. These characteristics limit the means that are practical for storing hydrogen, either in place or for transporting it. An examination of various means of storing hydrogen follows.

### Cryogenic

Cryogenic storage of hydrogen refers to the chilling of the element to very low temperatures to reduce it to a liquid state, reducing the volume. Hydrogen is a liquid at atmospheric pressure only at temperatures near absolute zero. Hydrogen is a liquid at temperatures below 20 Kelvin, which is the equivalent of -253°C or -423°F. Production of liquid hydrogen dates back to 1898 (Sir James Dewar). Cooling hydrogen to this temperature is also known as liquefaction, and the cooling process is energy intensive, consuming from 25% to 40% of the energy represented by the hydrogen stored. Current commercial operations lean toward the 40% figure.

The cryogenic vessels have to be extremely well insulated to prevent the hydrogen from returning to a gaseous state. The containment also has to include venting to prevent pressure build-up that could lead to explosions and which also permits cooling of the remaining hydrogen by evaporation. The addition of the safety gear, refrigeration coils, and extensive insulation adds to the bulk of the packaging. As a result, this technique is typically reserved for sizable quantities of hydrogen that will have to travel significant distances (typically a thousand kilometres or more—hundreds of miles).

Even with advanced cryogenic storage techniques, the rule of thumb is that there will be a 1% loss of hydrogen via evaporation or the equivalent in refrigeration energy consumed to maintain the volume, daily. Looked at another way, if you have a volume of cryogenically cold liquid hydrogen in industrial quality insulated containment, you will consume the equivalent of its energy content keeping it liquid in approximately 3 months. That's not a great shelf life.

Despite these issues, cryogenic storage is the preferred industry choice for transportation of hydrogen because it makes the hydrogen sufficiently dense to make the transport more cost effective.

However, even in liquid form at cryogenic temperatures, hydrogen is less energy dense than gasoline or ethanol; gasoline has 3.8 times as much energy as liquid hydrogen by volume.

There has been limited interest in using cryogenic storage for storing hydrogen on-board vehicles.

## Pressure

Automakers are typically using tanks to store hydrogen at pressures from 150 bar to 350 bar (roughly 2,000 to 5,000 psi) for their prototype hydrogen cars. However, this is not providing sufficient range for consumer acceptance, so now they are contemplating tanks rated to store hydrogen on-board at pressures up to 10,000 psi, or nearly 700 bar.

That's one of those numbers that doesn't mean much to most of us. Just what does 700 bar (10,000 psi) mean? Well, it's the same pressure as one would experience at a depth of over 7,000 metres (23,000 feet) below the ocean surface. It would crush any manned submarine yet built. This pressure is sufficiently high that steel is not considered strong enough to provide the containment, so more exotic tank designs (and materials) will be required. In addition, combining hydrogen-embrittled steel with very high-pressure hydrogen is a recipe for containment failure. Compressed air guns launch potatoes over 300 metres (350 yards) using 9 bar (130 psi), or about 1/80$^{th}$ the pressure in these proposed high-pressure containers. Steam locomotives used boilers that were typically rated for about 20 bar (300 psi), or 1/35$^{th}$ of these hydrogen tanks. Containment failures of such boilers were very ugly events.

Hydrogen storage at relatively low pressures dates back to the zeppelin era. Storage at 300 bar dates from the mid-1990s (Mercedes Benz NeCar 1 proto-type). It has not been used in any commercial application to date. Storage at 700 bar is quite recent and is essentially still experimental.

Due to the hazards associated with such high pressure, the vessels will have to be tested frequently to assure their integrity. Twenty-pound propane tanks that are designed for pressures up to about 20 bar (300 psi) must be re-certified every 10 years. The re-certification process is sufficiently expensive that most propane tanks are not re-certified, but simply destroyed after their initial certification period expires.

It takes energy to compress hydrogen. Current estimates indicate that com-pressing hydrogen to 700 bar at a commercial scale will take at least 15% of the value of the energy embodied in the compressed hydrogen.

While it may be feasible to build a tank that will safely store hydrogen at 700 bar, even in a high-speed collision, this also needs to apply to the filling connec-tion and the fittings, pipes, and hoses that will carry the hydrogen from the stor-age tank to other components in the vehicle. High-pressure hydrogen leaks have a special hazard associated with them—self-ignition. The jet of high-pressure hydrogen from a nozzle or a leak has the potential to generate a significant static electrical charge. The electrical potential is sufficient to generate a spark, right at the edge of the gas jet, which will be enough to ignite the jet if oxygen is present (as in the Earth's atmosphere).

700 bar might solve the on-board storage density issue for the automakers, but I don't think I want to be sitting in the same intersection with one of these tanks, let alone in the same vehicle.

## Chemical Carriers

One way to get around the issues of storing hydrogen in its pure form (low den-sity, gas, reactive, very leaky) is to allow it to be bound chemically to other ele-ments to make substances that are more convenient to handle and use. In general, this means a compound that contains a relatively high proportion of hydrogen, and one where extracting the hydrogen is relatively easy. This includes materials like ammonia, methane, ethanol, methanol, sodium borohydride (also known as sodium tetrahydroborate, sodium tetrahydridoborate) and others. Some of these have unpleasant characteristics in their own right (ammonia), or are still gases (methane).

In all cases, having the hydrogen bound to other elements means that the chemical carrier has to be cracked or reformed to release the hydrogen as required. The cracking process not only requires energy (reducing the overall system efficiency), it also requires additional equipment to be carried on board to carry out the reforming process. It also produces some kind of by-product as a result of the reforming. Still, the most successful fuel cell prototype vehicles to date in terms of range have typically used methanol (a liquid fuel at atmospheric pressure and ambient temperatures) as their on-board storage medium (e.g., NeCar 4).

Ethanol also shows great promise as a hydrogen carrier, and it too is a liquid fuel. Methanol carries four hydrogen atoms per carbon atom, where ethanol carries three hydrogen atoms per carbon atom. Methane, methanol, and ethanol were described in some detail earlier in the book (see the chapter on Hydrogen from Fossil Fuel Sources—Natural Gas and Methanol, and the chapter on Hydrogen from Sustainable Sources—Ethanol). In either case, carbon will either be released to the environment as these fuels are used (most likely), or have to be captured on-board to prevent release into the atmosphere (likely as carbon monoxide or carbon dioxide). While still cleaner than the combustion of gasoline or diesel fuel, this would tarnish the clean reputation of fuel cells as transportation technology.

Given the success of methanol reforming for use with hydrogen fuel cells, progress on research with fuel cells that can use methanol directly without reforming, and that methanol is a convenient fuel to produce, transport, and store, it should not be ruled out at this stage. As well, methanol and ethanol can be dispensed using the existing liquid fuel distribution infrastructure with minimal modifications, eliminating the need to invest in a new (hydrogen) fuel distribution infrastructure.

## Ammonia

Ammonia is a potent hydrogen carrier. Once cracked, its only by-product is nitrogen, which makes up 80% of our atmosphere. $NH_3$ carries three hydrogen atoms for each nitrogen atom, making it a very hydrogen-dense compound.

Unfortunately, it is rather nasty stuff for humans, being highly corrosive (alkaline) and a poison. It can be fatal if inhaled, and can be absorbed through the skin. It is typically a gas at atmospheric conditions, although it boils at -33.3°C or -28°F (which is why it is used for maintaining ice rinks). Given these properties, it simply should not be considered for use in a consumer application, especially in vehicles where collisions could lead to sudden releases of ammonia.

Ammonia is a common industrial chemical, and is used in some household cleaners in very diluted concentrations. It has been used in large volumes commercially for many decades.

## Sodium Borohydride

The best-known example of sodium borohydride used as a fuel is the Chrysler Town and Country Natrium concept vehicle (minivan), unveiled in 2002. The vehicle boasted a range of 480 km (300 miles) between refuelling stops, which is quite impressive amongst the fuel cell technology contestants. Sodium borohydride ($BNaH_4$) is made from borax, the stuff used for laundry detergents. It is a stable solid at atmospheric temperatures (up to 400°C), usually forming a powder or crystals. Hydrogen is extracted from the fuel, leaving sodium borate ($Na_2B_4O_7$, also known as borax, sodium borate decahydrate, or sodium pyroborate) as the spent fuel. Borax is also a solid, usually a powder or in flakes, at normal atmospheric conditions. This is collected on-board the vehicle, and exchanged for fresh sodium borohydride at refuelling time. The key elements are sodium (a common element and constituent of table salt—sodium-chloride) and boron, which is not found frequently in elemental form.

Admittedly, sodium borohydride has some handling issues, and is not a liquid fuel. Neither it nor borax should be ingested or inhaled. Still, based on my researches to date, this strikes me as the best solution for hydrogen fuel cells so far. So it seems odd that there is so little work being done on refining this approach as a solution. Perhaps it is because venturing into the world of borax brings us to consider elemental boron and its properties.

## Boron

If one wants to seriously visit the notion of boron derivatives as fuel (hydrogen carrier), one should also consider the potential of elemental boron as a fuel. Boron can take several forms, as a metal, powder, or crystal. The material is quite reactive, so it is conducive to electron movement, which is the basis for its use for energy storage (fuel). There is a compelling argument that boron would make a potent and very-low emissions combustion fuel, and with no need for fuel cells.[1]

---

1.    For a good overview of the boron energy cycle and the potential for this fuel, visit
      http://www.eagle.ca/~gcowan/boron_blast.html#TOC

## Hydrides

Hydrides currently store little hydrogen by weight relative to other options. The second major issue is the speed at which hydrides can be coaxed to absorb and release hydrogen, and finding methods that are less energy intensive to accomplish both operations (typically, cooling and heating are used now).

Most of the work on hydrides involves metals like magnesium (Mg), titanium (Ti), and lanthanum (La). These metals tend to be on the rare (expensive) side. Magnesium is almost never found in elemental form, so it has to be extracted and refined. It must also be coated for storage to prevent it combusting on contact with air. Titanium is valued for its strength and low weight, and likely will be more valued in other applications. It is also rare in elemental form, so has to be extracted and refined. Lanthanum is actually called a rare earth metal.

The operating temperatures of hydride storage are typically higher than the boiling point of water, which creates a risk factor when used in passenger vehicles.

Research on metal hydrides for the storage of hydrogen dates back to at least the 1960s. After approximately fifty years of research, it is still not a cost-effective solution for storing hydrogen. Still, R&D continues in hydrides despite the expensive materials, temperature of operation, efficiency issues, and weight of the storage material relative to the hydrogen stored because it is seen as a much safer alternative than cryogenic or high-pressure storage, and as having advantages over reforming the hydrogen from chemical carriers on-board the vehicle. Advances are being made on metal hydrides.

## Nanotubes

Of the various leading means of storing and transporting hydrogen, nanotubes are the newest. The technology is very new, and what little data is available is sketchy and variable, as the fundamental technology is still in the rapidly evolving research stages. Commercial scale operations do not exist yet, nor do reasonable cost estimates on its use.

The basic technology involves arranging carbon atoms into complicated structures at a molecular level that provide a storage matrix for hydrogen atoms. The hydrogen is adsorbed onto the carbon structures. Various techniques for forming the carbon structures and storing and extracting the hydrogen are still being developed and investigated. It will be some time before this approach reaches any degree of commercial maturity and standardization.

## Zeolites

This is a relatively new area of research as well for hydrogen storage and transport. Zeolites are either naturally occurring or manufactured crystalline lattices, which can be made of various materials. The zeolites are porous solids, which are frequently composed of materials like silicon and aluminum.

Zeolites are used in relatively small quantities in industry today, and tend to be expensive. The cost tends to limit their application to niche applications, primarily as catalysts. Their ability to adsorb specific gases while excluding others is the basis for interest in zeolites as a hydrogen storage medium.

# 9.2. Consumption (Conversion to Useful Energy)

## Fuel Cells

A significant number of organizations are working on various fuel cell technologies. I will not list the companies; they have their own PR groups to try to build up share value and support strategies to acquire additional funding. Besides, many of the players over the years have come and gone, so trying to keep track of them is a chore in itself.

There are multiple fuel cell technologies currently in use, or under development. A fuel cell is a device that consumes a fuel, producing electricity (and other by-products). The first way to categorize current fuel cells is by what they use as their fuel.

1.  Currently, most of the attention is being focused on fuel cells that consume hydrogen and oxygen (typically, from the atmosphere) to produce electricity, and usually water, heat, and some trace outputs that result from impurities in either the hydrogen or the oxygen, or degradation of catalysts used in the processing associated with the fuel cell.

2.  However, there are fuel cells being built today (and refined to increase capacity) that use ethanol or methanol as their fuels. These are worthy of more consideration because both fuels are liquid at ambient temperatures. Both are more energy dense (by volume) than hydrogen and both can be produced at a higher efficiency than hydrogen. A distribution infrastructure already exists that could easily accommodate these liquid fuels as transportation fuels. Ironically, there is also research going on for (direct) carbon fuel cells, which will actually produce the greenhouse gas carbon dioxide as a major by-product.

There are five major technologies that use hydrogen directly as a fuel. As you consider the following, remember that humans are comfortable at temperatures between 20° and 25°C (68° to 77°F), body temperature is 37°C (98°F), and temperatures above 50°C (122°F) cause scalding or burn injuries.

Only the solid oxide fuel cell can claim an efficiency barely exceeding 50% based on net, average electrical energy produced as a fraction of the energy embodied in the hydrogen consumed.

## Alkaline

Alkaline fuel cells operate at temperatures from 90° to 100°C (194° to 212°F).

Electrolytes can be poisoned by carbon dioxide, so typically require both pure hydrogen and pure oxygen to be supplied. More appropriate in environments other than the planet's surface; for example, underwater (such as submarines or submerged habitations) or outer space.

The typical electrolyte is potassium hydroxide (KOH) (also known as caustic potash). If KOH is at room temperature and it comes in contact with skin, it will cause burns. Platinum catalysts are used to enhance the reaction in these fuel cells.

## Molten Carbonate

Molten carbonate fuel cells operate at temperatures around 650°C (1,200°F).

They take a long time to reach operating temperature and to respond to changes in electrical demand. Therefore, their best application is in steady demand situations (e.g., utility base load).

The combination of materials (notably alkali metals and alkaline compounds) used and the high temperatures lead to high rates of corrosion and early failure of parts of the cell, or the entire cell.

## Phosphoric Acid

Phosphoric acid fuel cells operate at temperatures that range from 150° to 250°C (300° to 480°F).

The use of a platinum catalyst tends to make operations expensive. Phosphoric acid is highly corrosive.

Working units date back to the 1970s. While used primarily for mid-size stationary power generation, there has been some interest in this technology for transportation.

## Proton Exchange Membrane

Proton Exchange Membrane (PEM) fuel cells operate at temperatures around 80°C (175°F).

They don't like the cold (i.e., temperatures below freezing point of water, an output). PEM fuel cells are easily poisoned by even tiny quantities of impurities in the cell such as carbon monoxide or sulphur. Current primary issues are life, efficiency, and cost.

To date, the PEM technology has received serious attention only for the transport sector. This is likely because the PEM fuel cell has a better ability to change output levels quickly in response to changing demand than the other hydrogen fuel cell technologies. PEM fuel cells also reach reasonable output levels from off faster than the other types of fuel cells.

## Solid Oxide

Solid oxide fuel cells operate at temperatures around 800°C (1,500°F).

These cells take a long time to reach operating temperature and to respond to changes in electrical demand. Therefore, their best application is in steady demand situations (e.g., utility base load).

Fuel cells are not the only way to use hydrogen to produce useful energy to do work, e.g., power vehicles.

## Internal Combustion Engines

Automakers (notably BMW, and to a lesser extent Ford/Mazda and General Motors) have worked with hydrogen as a fuel for internal combustion engines at the R&D level for decades. However, producing engines with acceptable power, fuel consumption, and longevity has proven elusive. The requirement of engines that can run on either hydrogen or a fuel already in common use (e.g., gasoline), has complicated matters still further. The result is what we see today, no vehicles available to the public that can be fuelled with hydrogen.

Interest in hydrogen internal combustion engine vehicles (HICEV) may be increased due to the current California Air Resources Board (CARB) requirement to have 250 hydrogen-fuelled vehicles on the road by 2008. The Hummer H2 supplied to California Governor Schwarzenegger for his late 2004 photo-op was actually a HICEV, rather than a hydrogen fuel cell electric vehicle (HFCEV) as originally promised by the governor when running for office.

HICEVs produce emissions other than water vapour (e.g., nitrogen-oxides, a smog pre-cursor, and presumably some ammonia, a poison), as they combust air from the atmosphere rather than using (relatively) pure oxygen.

## 9.3. Hydrogen and the Consumer Society

Realistically, do we really want to deal with raw hydrogen? Nature doesn't.

Hydrogen is highly reactive; it's as if it wants to be anything else but free hydrogen. It seems to me that if someone really wants to make a mass-market fuel cell for in-vehicle use, it will not be based on using pure hydrogen. Instead, it would be based on using something easier to handle, store, and transport. I don't mean gasoline, on which DaimlerChrysler was wasting money a few years ago. I am thinking of a fuel cell that will use methanol or ethanol directly. A liquid fuel, that is easy to retrofit into the existing fuel structure. In fact, many fuel depots already dispense ethanol/gasoline blended fuels (aka gasohol). It has a high hydrogen content and low carbon content, and is possibly not as environmentally pure as a hydrogen fuel cell, but were folks really going to collect the water vapour exhaust to drink it anyway? I am aware of a couple of companies currently working on very small fuel cells that use methanol as their primary input for powering cell phones and laptops.

Hydrogen simply is not an attractive consumer product.

## 9.4. Still a Few Bugs to Work Out

In summary, the technology for storing and transporting hydrogen safely, effectively and in sufficient volume to meet the apparent needs of drivers is not ready. The fuel cell technology has not yet achieved basic targets for efficiency, robustness, or lifetime. Hydrogen combustion efficiency is even lower than that of fuel cells, which makes range highly problematic. On the whole, the hydrogen energy cycle as a practical device remains in the R&D stage, and is not ready for implementation.

Progress will be made on many of the outstanding issues related to the technology due to the vast resources being spent on them. However, we must be careful not to mistake progress in one small area for a complete, workable, and economically viable solution.

# 10

# *Hydrogen Cycle Efficiency*

Where the production of hydrogen is based on the use of fossil fuels, either as a feedstock, an energy source, or both, it actually takes considerably more energy to produce the hydrogen than is embodied in the hydrogen itself. Given that using fossil fuel energy and fossil fuel feedstock to produce hydrogen creates carbon and other waste products that are pollutants, how is the use of more energy supposed to provide us with the cleaner air promised in the 2003 State of the Union Address, or help us to reduce dependency on imported fossil fuels?

The industry that actually produces hydrogen from natural gas is not forthcoming with efficiency figures for the process. Based on what I have been able to glean from various sources, the figures vary depending on the quality of the natural gas source. However, a figure of 70% efficiency is likely reasonable (perhaps optimistic) as an average for the industry. Other research has used figures ranging from 50% to 75% as the overall efficiency of industrial processes to produce hydrogen from natural gas[1]. That is, the hydrogen produced from the process embodies less than 70% of the energy that was available in the natural gas (and conceivably other energy sources) used to produce the hydrogen. This figure does not include any penalty that might be associated with the recovery or capture of the carbon and other impurities that are by-products of the process, or sequestering the carbon.

## *10.1. Case Study*

I am going to start from what I consider the optimal case for hydrogen as an environmental transport fuel. Wind turbines in areas remote from major urban cen-

---

1. Consoni, Stefano and Viganò, Federico, paper presented at Second Annual Conference on Carbon Sequestration, Washington D.C., May 5-8, 2003 http://www.princeton.edu/~cmi/research/Capture/Presentations/consonni.de-carbonized%20hydrogen.ppt (slide 5)

tres will produce electricity intermittently, but relatively steadily over periods of weeks. This electricity will be used exclusively to produce hydrogen from water via electrolysis, avoiding the issues of dispatchable power, backup generation facilities, and increasing cost of the facilities. I am assuming exclusivity so as to avoid any costs associated with building new transmission lines. Besides, if the facility was grid-connected, there would be no surplus power to be used for generating hydrogen—the amount of wind-generated electricity in North America will remain so small for decades to come that it could not meet even base load demand.

Commercial electrolysis operations appear to top out at about 75% efficiency, based on a quick review of literature on the subject. 60% seems to be more realistic in real-world operations. Assuming a generous 75% efficiency, the cost is a 25% energy loss as a result of the conversion from electricity to hydrogen. We are going to liquefy the hydrogen, for our example, to make it compact to transport because liquid hydrogen is the most compact storage form. It is also the prevalent form used by today's commercial hydrogen industry for transport. Liquefying hydrogen from room temperature uses approximately the equivalent of 40% of the energy stored in the hydrogen (including some leakage and evaporation losses). Let's simplify our calculations by simply assuming that only 60% of the hydrogen we put into the liquefying process appears as liquid hydrogen.

A special truck, an HFCEV, transports the liquid hydrogen from the production site to the distribution depot. It consumes as little as 4% of the energy it is carrying making the round trip each day. Keep in mind this figure is likely generous; pipelines and electrical transmission lines are typically considered to have losses in the range of 5% to 8% depending on the distance, and industry prefers pipelines to trucks when they have the option.

We'll assume the liquid hydrogen is transferred to the storage facility with no losses. However, keeping the storage facility cooled to 20 Kelvin is estimated to cost 1% of the volume per day. We will assume that 3 days' supply is retained on average in storage, which costs an additional 3%. The hydrogen is pumped from storage into on-board fuel tanks in each vehicle—the tanks may be cryogenic, but probably are not. We will generously assume zero losses here as well. Once in the vehicle, which is fuelled on average once a week, there is another 3% loss in storage due to continued refrigeration or warming/evaporation losses.

The hydrogen is next fed to a fuel cell. We'll assume the fuel cell has 40% net efficiency. Robust, commercial fuel cells capable of use in road vehicles have not reached this figure yet, and some look upon it as an unreachable, theoretical figure. Larger, non-mobile fuel cells are now achieving 35% net efficiencies under ideal operating conditions, but not continuously. This seems to be the deep, dirty secret of the HFCEV industry. They tout the 90-plus per cent efficiencies of the electric drive train, but are curiously silent about the efficiency of the fuel cell itself. Fuel cell statistics are almost always presented in terms of kW of capacity as peak power output, not efficiency. The nascent HFCEV industry points out that the fuel cell plus electric drive train is more efficient than an internal combustion engine, but conveniently ignores efficiency comparisons with battery electrics. Still, we'll give the hydrogen fuel cell every break we can in this analysis and use the 40% figure.

The fuel cell is not sized to provide peak power for the vehicle, but something just over typical operating power demand. Super-capacitors or high-power batteries are used as a buffer to even out demand on the fuel cell, and this charging and discharging consumes some power, which I am assuming to be about 2%. Only now do we have the equivalent of the power that comes from batteries in a battery-only electric vehicle (BEV). So let's recapitulate all those losses, in terms of efficiencies, starting from the power from the wind turbine.

| Process | Efficiency | Loss |
| --- | --- | --- |
| Electrolysis | 75% | 25% |
| Liquefaction | 60% | 40% |
| Transport | 96% | 4% |
| Bulk storage | 97% | 3% |
| Vehicle storage | 97% | 3% |
| Fuel cell | 40% | 60% |
| Super-capacitor | 98% | 2% |

How does that stack up? Well, when we do the math (75% x 60% x 96% x 97% x 97% x 40% x 98%=15.93%), we get a total efficiency of under 16%. Remember, this number is undoubtedly optimistic, given the generous assumptions we have used in developing the scenario. Other estimates have provided figures as low as 4% cycle efficiency. So while we can quibble with one or more of the specific numbers used here, it won't change the essential conclusion: the overall cycle efficiency of hydrogen fuel is unacceptably low compared to technology alternatives already available and in use.

If that same kilowatt-hour (kWh) of electricity that was originally produced by the wind turbine was transmitted via high tension lines and the local distribution grid (8% loss) and used to charge lead-acid batteries (10% loss, including one day's worth of self-discharge loss—approximately 1%), the equivalent efficiency (from turbine to motor controller) would be 83% efficient; more than 5 times better than the hydrogen storage and transport cycle.

I have been making pretty generous assumptions all along to the benefit of the hydrogen fuel cell energy cycle, so the 16% figure is again undoubtedly generous. It still ignores the cost of building the electrolysis facility, the transport vehicle, the bulk storage facility, and the fuel cell itself. I am assuming that the advanced batteries and the on-board hydrogen storage, if other than low-pressure gas, will have similar prices for the foreseeable future. This analysis has also ignored the remaining R&D remaining for the safe and economical handling, storage, transport, and dispensing of hydrogen.

Is it worth producing five times as much electricity to travel a kilometre in an HFCEV, than the same distance in a BEV? Before answering that question, remember that more electricity is produced in North America from burning coal or natural gas than from all sustainable power sources combined.

There are those who will argue that the efficiency of fuel cells will improve, and exceed even the 40% figure I have granted in the analysis above. That could be, but fuel cells will never reach the over 300% efficiency they would have to achieve to match the energy efficiency of the BEV already on the road today. 300% efficiency means you get three times as much energy out of a device as you put in. That's over-unity, which is impossible.

In terms of creating an efficient energy system, making hydrogen is a losing proposition.

Where the primary energy to produce hydrogen comes from clean, sustainable sources, the losses associated with converting that energy (typically electricity to run electrolysis processes) to hydrogen for subsequent use (often to produce electricity again) are substantial.

In our current situation, where sustainable energy sources make up such a small part of our energy supply, it is folly indeed to find ways to use that energy even less efficiently than we are doing now.

Not only is hydrogen not an energy source, it is not even an attractive energy carrier.

A paper by Stephen  and James Eaves[2] confirms my essential conclusions regarding the efficiency of the hydrogen energy cycle, especially as compared to

---

2.    *Cost Comparison of Fuel-Cell and Battery Electric Vehicles*, Eaves, Stephen and Eaves, James, Current Events magazine, September-October 2004

EVs with advanced batteries. They have made even more generous assumptions than I did regarding hydrogen fuel transport, but still conclude the HFCEV will consume at least 2.4 times as much primary energy as the advanced BEV (ABEV) to perform equivalent functions.

# 11

# *The Proposed Hydrogen Timeline*

The timeline below has been constructed from dates provided by government agencies, credible authorities, and hydrogen proponents in press releases and other documents they have placed in the public domain.

2008      Automakers are required by California Air Resources Board to have 250 vehicles running on hydrogen in North America. They do not have to be available for sale, or use fuel cells. This condition was in exchange for killing the California EV mandate, which had succeeded in producing success stories like the GM EV-1 and Honda EV+ and other capable on-road, zero-emissions vehicles.

2008      Kyoto Convention on Greenhouse Gas Emissions Reductions starts its phased-in implementation, as sufficient countries have ratified it as of 2004. The U.S. has not ratified the protocol, and indicates no intention to do so.

2010      California Hydrogen Highway to be completed with almost 200 hydrogen filling stations spaced at intervals of no more than 20 miles. As of 2004, standards have not been set for the means of distributing the fuel (liquid or pressurized gas), although the first was scheduled to be in operation before the end of 2004 (but not open to the public). The executive order to start this work was signed into force by California Governor Arnold Schwarzenegger on April 20, 2004, with an estimated cost of US $100,000,000.00. The Governor's Office has stated on several occasions that hydrogen cars will

|      |      |
|------|------|
|      | be available to the public by 2010. No automaker has yet stepped up to that date. |
| 2012 | The Kyoto Convention on Greenhouse Gas Emissions Reductions comes into full force. |
| 2012 | The earliest possible date DaimlerChrysler claims they will have hydrogen-fuelled vehicles available for sale, even in limited numbers for fleet sales (per announcement in March 2005). DaimlerChrysler is considered a leader among automakers on hydrogen fuel cell vehicles. (In 1999, DaimlerChrysler and other automakers predicted they would have hydrogen vehicles in limited production by 2004.) |
| 2014 | The earliest possible date Toyota or Nissan believe they could have hydrogen fuelled vehicles available for sale, even in limited numbers (per announcements in December 2003). |
| 2015 | Technology to be sufficiently advanced to permit industry decision regarding viability of commercialization in the U.S. (per U.S. DOE in 2004 "Hydrogen Posture Plan") |
| 2020 | According to the U.S. White House (February 6, 2003): "Through partnerships with the private sector, the hydrogen fuel initiative and FreedomCAR will make it practical and cost-effective for large numbers of Americans to choose to use clean, hydrogen fuel cell vehicles by 2020." http://www.whitehouse.gov/news/releases/2003/02/20030206-2.html |
| 2020 | Estimate of when worldwide discoveries of natural gas (the primary source for hydrogen production today) will peak. North American discovery peak has already passed, and production has plateaued, and may be in early stages of decline as of 2005. |
| 2025 | The earliest that fuel-cell vehicles will be in significant production—1,000,000 vehicles per year or more |

(according to Arlington Institute forecast in 2003, based on a "moon shot" level of effort, and assuming first fuel-cell vehicles are available to the public in 2012). Conservative estimates suggest that building the necessary infrastructure to produce the hydrogen and distribute it in the U.S. alone will cost over US $100,000,000,000.00 (one hundred billion dollars). That doesn't include the vehicles, or primary energy sources.

2030    Weyburn Carbon Dioxide Sequestration monitoring project scheduled to end.

2040    U.S. DOE estimates 150 megatons of hydrogen will be consumed annually to fuel the growing fleet of cars and light trucks powered by hydrogen in the U.S. That's 16 times the total industrial production in the U.S. for 2003.

2040    The optimistic target for Iceland to embrace the hydrogen economy (according to Ingimundur Sigfusson, Iceland's ambassador to Japan). Iceland is widely regarded as the world leader in moving to the hydrogen economy.

2050    The year in which the United States National Research Council of the National Academies thinks hydrogen could actually be in place as a major fuel in the U.S. http://www.whitehouse.gov/news/releases/2003/02/20030206-2.html

Of course, there are no guarantees that any of the above will actually come to pass, and plenty of reason for scepticism. The dates listed above are provided by the hydrogen economy's advocates and supporters. However, even if we put on our rose-tinted spectacles and embrace this euphoric vision of the future, a simple question remains.

If the problems the hydrogen economy is purported to solve are truly real and important enough to justify this magnitude of effort and investment, can we afford to wait this long?

Given reasonable assumptions about our future energy use, our remaining fossil fuel reserves, and our lack of action on developing sustainable energy resources, and even granting the optimistic assumptions set out above, the hydrogen economy can't be in place in time to fill the gap when oil becomes too expensive for industrialized countries to continue with business as usual.

# 12

# *Hydrogen Cars as Power Sources*

Several authors have posited a scenario where hydrogen fuelled cars will be ubiquitous. Ballard states "fuel-cell cars are poised to be our next enduring love."[1] President George W. Bush said, "with a new national commitment, our scientists and engineers will overcome obstacles to taking these cars from laboratory to showroom, so that the first car driven by a child born today could be powered by hydrogen and pollution-free" in his 2003 State of the Union Address.[2] Others have gone further, and see these hydrogen cars morphing into mobile distributed generation power sources. Jeremy Rifkin wrote "in the new hydrogen fuel-cell era, the automobile itself is 'a power plant on wheels' with a generating capacity of twenty kilowatts".[3]

In short, it goes like this. The hydrogen-fuelled vehicle will be refuelled either at the owner's home while parked or at a commercial refuelling depot. Then, when the owner parks the car (at work, at home, while running errands), she or he will plug the car into an intelligent outlet, which will communicate with the car and the power grid—this is becoming known as vehicle to grid (V2G).

When the power utilities need additional energy, they will poll cars (and other distributed generation sources plugged into the grid) if they are prepared to provide power at a specified price (auction). Each owner is expected to program her or his vehicle with an acceptable price threshold, and how much power she or he is prepared to sell from the reserve on-board the vehicle (e.g., do not deplete below 50% of storage capacity).

There are a few issues with this picture. There are some logistical issues that have to be overcome, notably establishing a communications standard (both physical and protocols) that will be used by both the power industry and car manufacturers. It is feasible, but a significant piece of work.

---

1.  *Hydricity, The Universal Economy*, Ballard, Geoffrey, in *Fueling the Future*
2.  2003 State of the Union Address
3.  *The Hydrogen Economy*, Rifkin, Jeremy, p. 208

The power utilities today are very reluctant to allow resources they do not own and control to supply power onto the grid, generally known as back-feeding. According to the power companies, this is to safeguard their workers from shocks when working on sections of power lines that they expect to be un-powered when doing maintenance. Even if the workers check the section for power before starting work, they cannot be guaranteed that power from an uncontrolled, interconnected source will not be applied while they are doing the work. This will be a greater concern for the utilities where the temporary power sources are mobile and unscheduled.

It is already becoming a matter of some concern in connection with hybrid vehicles. According to Mark Duvall, EPRI's manager of technology development for electric transportation, "if you asked 20 different utilities today what they thought of vehicles putting power back into the grid, you wouldn't get a very positive response. It took a long time to assure the utility industry that it was worthwhile just to plug solar and other items into the grid. It's going to make them very nervous".[4] EPRI is the Electric Power Research Institute, based in the United States.

The conventional North American electrical outlet is designed to carry approximately 1.5 kilowatts (kW). A typical household service is designed to carry up to 24 kW. The typical hydrogen fuel cell powered car will be capable of delivering at least 100 kW (and possibly double that) to provide acceleration performance suitable for North American highway driving. It may be the fuel cell will not be sized to produce that much power, but may rely on batteries to provide peak power for acceleration. In this case the fuel cell would be sized to meet average demand or a bit more in a fuel cell-battery electric hybrid. This would be more in the order of 15 to 25 kW according to the size of the vehicle. However, to date the would-be fuel cell carmakers have essentially stopped work on developing appropriate buffer battery technologies.

Significant upgrading of the retail end of the power grid will be necessary to support the power capabilities of the mobile fuel cells, and to provide the intelligence necessary to control and administer it. Such upgrading of the retail end of the electrical grid is a major expense the electrical utility industry has been trying to avoid for decades.

The administration costs will be significant. Each vehicle will require a unique ID in order to support compensation. The vehicle ID will have to be matched to an owner, to receive the compensation. Each outlet (not just at the circuit or ser-

---

4.   *Electric Cars That Pay*, Clayton, Mark, Christian Science Monitor, July 29, 2004

vice owner level) will have to have an ID, as multiple vehicles could be attached to a single circuit via multiple outlets. Monitoring will have to be applied at the outlet level to provide data for subsequent compensation, and to ensure that the total power supplied does not exceed the rating for the complete circuit. The outlets will have to be labelled as to who owns them, so it can be determined which power company received the power, and is responsible for paying for it. Accounts will have to be maintained, and payments issued. This administration will entail a sizable initial investment and operating overhead cost.

The province of Ontario is currently estimating the cost of Smart Meters for household service (for simple one-way power delivery, and not two-way power flow or net metering) at Cdn $100 per unit in high volumes. The devices that would have to be installed at each outlet to support V2G connections will have to be even smarter (peer-to-peer as well as master-slave communications) and presumably more expensive.

Each vehicle will require the ability to provide the power, not as raw DC electricity at the fuel cell's native voltage, but instead as 60-cycle AC electricity. As well, this power must be synchronized to the grid, at the specified voltage, which may vary anywhere from 100 to 250 volts depending on the type of connection used and the current loading on the grid. The current will also have to be limited to what the circuit can safely support, and the circuit may not be dedicated to this single power source at a given time (e.g., more than one vehicle could be connected to a circuit in a parking garage). This capability will add a significant cost to each vehicle produced; in addition to the cost of having the intelligence on-board to monitor current fuel status, and communicate with the grid both at the auction (financial) and power supply (physical) levels. The vehicle will also have to maintain an independent record of the amount of power provided, including dates, times, and agreed rates, for subsequent reconciliation of accounts with the power company's records. Presumably, the vehicle data logging system will also have to be able to transfer data to other computers for aggregation and analysis of the data and reconciliation with the statements from the utilities that received the power.

Advocates of the V2G have suggested that selling the power from vehicles to the grid will be so lucrative that it will "defray the cost of the lease or purchase price of the vehicle."[5] But will it really pay off for the vehicle owner who pays a premium for such a vehicle?

---

5.    *The Hydrogen Economy*, Rifkin, Jeremy, p. 208

Let us assume the power a hydrogen fuel cell vehicle can supply to the grid is limited by the circuit to which it is connected, which is rated at 20 amperes and 120 volts. The vehicle will be able to provide approximately 2 kW continuously to the grid. If the utility is prepared to offer $0.15 per kWh to vehicle owners (to avoid paying a higher cost to purchase it elsewhere at peak demand periods—effectively their wholesale cost), and the peak demand period extends for 4 hours a day and the vehicle is available for that entire period, the owner may be able to earn an entire $1.20 for the day. In the unlikely event that this can be maintained for 20 days a month, the vehicle owner can earn up to $24.00 for the month. That's probably a maximum, and the utility is unlikely to pay the full rate for vehicle-based power, as it cannot be guaranteed that the vehicle will be available for the entire time it needs the additional power to meet the excess demand. At these power levels, the cost-benefit ratio for this investment is not likely to be attractive to the utilities.

Remember, the electrical utility is in the business of producing or buying electricity at one price, and selling it at a higher price. The margin is what pays for its other costs (e.g., maintenance, salaries) and provides its profit.

However, to make this possibility more attractive to both parties, let us assume the vehicle can be connected to a higher capacity connection, say up to 25 kW. State of the art hydrogen prototype vehicles are currently carrying enough hydrogen to provide approximately 25 kWh of electricity from their fuel cells (at current state of the art fuel cell efficiencies). So the vehicle owner can now provide 25 kW to the grid for an hour. At $0.15 per kWh, they will have earned $3.75. However, their hydrogen tank will be empty and the vehicle will have to be towed home.

The fact is, most of the current hydrogen fuel cell prototype cars are averaging less than 100 miles between refuelling stops, and it is unlikely many vehicle owners will be willing to give up much of that range to earn a couple of dollars from the local electrical utility on a daily basis. However, it is possible that larger, safe storage capabilities will become feasible with further R&D. After all, Kordesch was achieving over 300 km between refills in the early 1970s.

There is also the matter of basic efficiency, for those who care about such things. Suppose the initial hydrogen on-board the vehicle is produced using electrolysis via a home-based electrolysis unit. The initial electricity purchased or provided by the homeowner will be used to produce hydrogen, oxygen, and heat, and the resulting hydrogen will embody approximately 50% of the energy in the electricity that was used to produce it and compress it so that it could be stored on board the vehicle. Subsequently, this hydrogen will be transformed back into electricity using a fuel cell that resides in the vehicle, with a maximum net efficiency of 50%, probably significantly less. Ignoring other minor losses (e.g., evaporation losses, transmission and line losses, more realistic conversion efficiency factors), the best case for producing electricity to meet peak demand via this method is to retrieve just one-quarter (25%) of the electricity that was used originally to produce the hydrogen, possibly just hours earlier.

If hydrogen fuel cells are such a good idea for meeting peak demand requirements for the electrical utilities, the utilities are more likely to buy larger, stationary fuel cells that they can control than to invest in a massive distributed infrastructure to buy power from privately owned vehicles that may not always be available. Large, stationary fuel cells will likely be more efficient than mobile fuel cells, and cost less per kW of capacity to build.

If hydrogen fuel cells are going to be used to power the grid, running it from parked cars with limited on-board storage capacity is unlikely to be as attractive as using facilities that are designed for the purpose.

There is also the question of whether or not the amount earned by selling this power to the utility will cover the vehicle owner's actual fuel cost. Remember, the private vehicle owner is going to be paying for hydrogen at retail prices (either at the fuelling station or via the electricity used in a home-based electrolyzer); the electrical utility will be able to buy its hydrogen at wholesale prices.

In reality, electrical utilities are unlikely to use an energy storage mechanism that only delivers approximately 25% round trip efficiency unless the pricing is very attractive, i.e., subsidized by the vehicle owners. They have better options at their disposal (see the chapter on energy storage).

While powering the grid from hydrogen vehicles seems unlikely on a significant scale, the potential for replacing the grid using vehicle-based power in specific situations does hold some real promise.

Hybrids in the near term are expected to start providing this very capability. Major automakers have announced plans for "Contractor's Edition" light trucks that will use the on-board generators that are part of the drive train to supply power for electrical tools remotely from the grid. The ability to power homes from our vehicles will have a small niche market primarily in off-grid situations. These include utility trouble trucks, off-grid homes, and when the grid goes off. The latter is becoming increasingly common, both on the large scale (August 2003 blackout) and the small scale (rotating blackouts such as in California and Alberta in the summer of 2006 due to power demand exceeding supply during heat waves, as well as the more typical local outages due to equipment failure and damage).

# 13

# *Hydrogen Cars vs. Battery Electric Cars*

While you wouldn't know it from the press materials, Web sites, advertising, and shiny brochures, hydrogen fuel cell vehicles (HFCVs) are actually just electric cars with a fancy battery. The hydrogen fuel cells are simply another means of producing electricity to power an electric motor (or motors) to drive the wheels. So, when the hydrogen proponents praise the high efficiency of their drive trains (better than 80%), they are really paying homage to the venerable, quiet, zero-emissions, efficient, electric motor.

The HFCV is really the HFCEV: the hydrogen fuel cell *electric* vehicle.

This point raises a question blithely ignored by HFCEV advocates. Why will HFCEVs succeed where battery electric vehicles (BEVs) have failed? How do HFCEVs stack up against BEVs, or more importantly, against what advanced battery electric vehicles (ABEVs) can now achieve?

In general, it is hard to compare an HFCEV to a BEV because the vehicles used have been so different and so much depends on details of the design and construction.

Fortunately, there is one example of an HFCEV and a BEV that were built on the same chassis that could provide a reasonable comparison. The Honda EV+ minivan was produced in response to the California zero-emissions vehicle (ZEV) mandate in the 1990s. Although it was available through regular Honda dealer-ships, there were waiting lists as demand exceeded supply, so they did not sit in the showrooms. While this vehicle was available for sale at a manufacturer's sug-gested retail price (MSRP) of US $53,000, most were leased—as the US $433 monthly lease payment was very attractive by comparison (36 month lease). These leased vehicles were subsequently reclaimed by Honda, and some were re-used as the platform for the Honda FCX HFCEV prototypes. The Honda EV+ with early NiMH batteries had a range of up to 215 km (135 miles) per charge, and a curb weight of approximately 1633 kg (3594 lb). With access to distributed charging facilities, operators demonstrated they could achieve over 320 km (200 miles) per day with these vehicles.

Honda introduced its HFCEV, the Honda FCX, approximately a decade later. To build the FCX, Honda incorporated improvements resulting from the EV+ experiment. These FCX prototypes have a curb weight of 1700 kg (3740 lb), which is heavier than the BEV predecessor, and obtain a range between refuelling stops of about 240 km (150 miles). Honda officials advised not going beyond 200 km (120 miles) between refuelling stops. Still it's not an apples-to-apples comparison. The EV+ was actually in limited production and available to the public. Honda FCX fuel cell vehicles are not available to the public. Their estimated cost is still over $1,000,000 per unit. The fuel cell alone costs more than the 1990s retail price of the EV+ BEV. What would be truly interesting would be to see how this same platform would perform with a current technology advanced battery pack (e.g., lithium ion), sized for range and incorporating the benefits the FCX has gained as a result of the original EV+ lease program.

The Honda FCX is about as mass-produced as we have yet seen in an HFCEV. Honda expects to build up to thirty of these vehicles. They are not available for sale; to date no HFCEV has been made available for purchase. A small number (up to five) of the vehicles are being leased to the City of Los Angeles at US $500 per month, each. This bears no relation to the actual cost of the vehicles—conservative estimates in several media reports are that the average HFCEV (not only the FCX) produced to date costs well over US $1,000,000 apiece.

Although the FCX is considered state of the art, cold weather testing of the fuel cell vehicles has just begun (at temperatures slightly below 0°C or 32°F). In large parts of North America, much colder temperatures are a regular feature of normal usage. The FCX does not show appreciably better range than the 1967 GM ElectroVan hydrogen fuel cell prototype vehicle—40-year-old technology.

The 2004 FCX model claims a 15% increase in range, using hydrogen stored at 350 bar (5,000 psi). The NiMH EVs were also improving their capabilities year over year, before they were removed from the market.

In reality, the Honda FCX is a hydrogen fuel cell/battery electric hybrid. It uses an ultracapacitor (essentially a high-power, low capacity battery) to store electricity from the fuel cell when power demand is low and to deliver more power than the fuel cell can when power demand is high.

Other than the Honda EV+/FCX, it's difficult to find any example of an HFCEV and BEV on the same chassis (let alone an ABEV) to make the analysis simple. Even if it were possible to compare technical capabilities at this level, it is impossible to get real cost and pricing information from automakers regarding the HFCEVs, and the information provided to date on BEVs made by the automakers in the past decade appears suspect (it defies common sense).

Toyota sold the RAV4 EV in very limited numbers in California; well under a thousand. These were not mass-produced, but were essentially hand-built conversions, so economies of scale we typically expect from the automakers did not apply. The RAV4 EV was sold in California irregularly from 1999 to 2002, mostly to fleet operators, until the CARB ZEV mandate was effectively broken. The RAV4 EV had a MSRP of US $42,000. It appears there was no discounting on this price, as demand exceeded supply for the entire time the vehicle was offered, and a waiting list existed from the time the vehicle was announced until it was withdrawn from the market.

That sticker price may seem high to most U.S. car-buyers, and it was. It was sold primarily to a specific market that could afford the price and needed the technology for public relations purposes at least as much as for real, on-road use. However, several people who analyzed the vehicle felt this figure was excessive for what the vehicle contained. By comparison, the gasoline-powered version of the same chassis had a MSRP of US $17,000, but typically sold for less. Aside from the batteries, the electric components in the RAV4 EV were conventional. To further confuse the pricing issue, the effective price of the RAV4 EV to consumers was actually US $30,000 (after the CARB incentive of US $9,000 and federal tax credit of US $3,000), and this included the installation of a proprietary charger at the home of the vehicle owner, likely worth another US $2,000 to US $3,000. Income tax deductions at federal and state levels would reduce the effective price still further for most purchasers (depreciation as business expense). The MSRP undoubtedly took all these factors into account. The RAV4 EV is actually simpler than the Toyota Prius hybrid, which retails for approximately US $20,000. To my mind, this collection of figures suggests that such a vehicle, produced in reasonable quantities (tens of thousands, not just hundreds), should retail for no more than US $25,000, probably less (again, note the more complex hybrid retails for US $20,000).

The RAV4 EV had an advertised range of up to 160 km (100 miles) per charge with the smallish advanced technology battery pack offered. The similarly sized Honda EV+ managed a range of over 210 km (135 miles) per charge.

GM's EV-1 with NiMH battery technology got over 240 km (150 miles) per charge. However, even this was hardly pushing the envelope. It was more of an exercise in being just not quite good enough.

Solectria built a prototype vehicle called the Sunrise, which achieved a distance of 600 km (373 miles) on one charge in 1996 using an earlier version of this battery technology. The NiMH battery technology has advanced since then. Remember, this is early 1990s off-the-shelf technology, produced in the low thousands of vehicles, by at least four auto manufacturers, not battery makers. Significantly better battery technologies are available now.

Nissan has also introduced a prototype HFCEV, although in smaller numbers than the Honda FCX. The X-Trail FCV boasts even better range figures than the Honda FCX, largely because it is even more of an EV than the FCX. The X-Trail FCV boasts a significant lithium-ion battery pack to smooth out the demands on the fuel cell.

Will the costs of HFCEVs come down with development and increased production volume? Undoubtedly, assuming they get that far. Remember, a vehicle-based fuel cell capable of operation in truly cold weather with an acceptable life span has not yet been produced, let alone proven. Hopes for a version that can be mass-produced with acceptable quality and cost only comes after that. Will HFCEVs ever be cost-competitive with ABEVs? It seems very unlikely given the relative costs of components required (there's a lot more gadgetry in a battery or ultracapacitor/fuel cell hybrid vehicle than in a BEV). Will the cost of hydrogen as a fuel ever be competitive with simple recharging from an electrical outlet? Again, very unlikely, as the premier means of producing hydrogen so pure that it does not poison the fuel cells in short order is based on using electricity, and suffers two sets of significant conversion losses from there.

In December 2002, Honda and Nissan introduced their HFCEVs in California with great mass-media fanfare. Lost in the small print was the fact that these companies see mass-marketing of such vehicles as being at least another decade away. Honda announced it would lease a limited number of 2003 Honda FCX vehicles in the U.S and Japan by the end of 2005. It appears that did not happen. Instead, a revised version of the FCX concept vehicle was announced late in 2005, and in a press release in 2006, Honda announced that it would begin production of an HFCEV in 2009 or later.[1] Honda currently has no plans for mass-market sales of these vehicles. Fuelling will be based on reforming of natural gas at dedicated home energy stations, over an extended period of time (hours). Sounds a bit like recharging a BEV at home, doesn't it? Again, where's the HFCEV's advantage?

---

1.    http://world.honda.com/news/2006/4060108FCX/

At the 2002 publicity event, Nissan officials stated they only intended to start public road testing of the zero-emissions X-Trail FCV in Japan in 2003, with subsequent very low-volume production in early 2005. As of late 2005, there is no further word from Nissan regarding the X-Trail FCV on their Web page dedicated to the subject.

To date, the only way HFCEVs have managed ranges significantly better than off-the-shelf advanced batteries in limited production BEVs is to use 700 bar (10,000 psi) storage tanks, which are not yet certified as safe for road use. Even so, they are still not in the same league as the 1996 NiMH Solectria Sunrise. The ABEV has an equivalent option to higher pressure hydrogen storage when cost is no object—just add more batteries.

Is it really a good idea to let EVs be put on hold for a decade or more while we wait to see whether the fuel cell, and the HFCEV in particular, pans out? Or, is there a reasonable chance that the HFCEV is yet another panacea that will never become a market reality, and end up as another automotive industry stall tactic like the United States Advanced Battery Consortium (USABC) or the Partnership for a New Generation Vehicle (PNGV)? I don't think it is time to rest and believe, that this time, the automakers really mean to deliver a clean-air vehicle out of the goodness of their hearts. Note the lack of economic information that is currently being provided by the promoters of HFCEVs, which is how they eventually strangled the recent reincarnation of BEVs pried out of the automakers by CARB in the 1990s (e.g., the GM EV-1, the Ford Ranger EV, the Chrysler EPIC, the Honda EV+, the Nissan Altra, and the Toyota E-com). In fact, at the January, 2003 Detroit Auto Show, the big three North American automakers were already backing away from fuel cells and going back to hybrids as their PR strategy (in reality, electric-assist gasoline vehicles, not serious hybrids with robust electric drive-trains).

Early in 2006, Honda announced an improved version of the FCX that claims a range of over 500 km using a combination of metal hydride and high-pressure storage. However, if we are betting the future on press announcements, we should also note that Subaru announced a high-capacity lithium battery for its R1e prototype ABEV that can travel 120 km on a very small battery pack, can be recharged in under five minutes, and can presumably use a larger battery pack to provide increased range. Clearly, the PR competition between hydrogen and advanced batteries for range between refuelling stops is not over.

The HFCEV will use at least four times as much energy per mile (starting from electricity to either produce hydrogen or charge batteries) as the BEV.

The world's automakers have consistently said there is no market for BEVs because of their limited range on a charge. Yet they expect us to believe consumers will flock to the HFCEV with comparable range. To date, many hydrogen vehicle prototypes (HFCEV and HICEV—hydrogen internal combustion engine vehicles) are only marginally better than BEVs using lead-acid batteries, a long proven technology that is very economical and available off-the-shelf today. Only the state-of-the-art HFCEVs using very high pressure storage have autonomy comparable with decade-old advanced battery technology. Nor are they appreciably better than what was accomplished in the 1960s and 1970s with high-pressure hydrogen storage. To date, these are not great advances in autonomy over what was done four decades ago. None of the hydrogen vehicle prototypes have capabilities that compare favourably with the high end of the current generation of advanced batteries (e.g., lithium ion, lithium metal polymer).

Let's review the discussion. Hydrogen proponents, including the automakers, have effectively killed the production BEV. The BEV has the following advantages relative to the HFCEV (and HICEV).

- The BEV is less expensive to build, leading to a lower initial purchase price.

- The BEV uses less primary energy per kilometre (or mile) travelled (by a factor of four or more).

- BEVs are a mature, robust technology, available off-the-shelf today. Thousands are in on-road use today, millions are in use in off-road applications. HFCEVs and HICEVs are still in the prototype research and development phase.

- The refuelling infrastructure for BEVs is already in place and ubiquitous; at any electrical outlet. The type of refuelling infrastructure to be used for HFCEVs and HICEVs has not yet been agreed upon, and will cost billions of dollars to implement.

- BEVs are truly zero-emissions at point of use. HFCEVs will emit water as an exhaust product, which could be a problem on roads where temperatures are below the freezing point for water. HICEVs will produce additional emissions (e.g., $NO_X$).

- There is actually a surplus of electricity in North America at off-peak periods (e.g., midnight to 6 am), when most BEVs would likely be recharged (to take advantage of lower electricity prices) and owners are sleeping, not driving. Hydrogen is produced primarily from steam reforming of natural gas, an increasingly expensive fossil fuel.

Yet the world's automakers and the rest of the hydrogen herd would have us believe that hydrogen vehicles will succeed, someday, where BEVs have not. Why?

Why are the U.S. government (and other governments), the world's automakers, and the rest of the hydrogen lobby betting on an inferior technology (lower range, efficiency, and higher cost than ABEVs) that they freely admit won't be ready for at least another decade (probably longer), instead of a superior technology that is ready for use today?

This is especially curious, given that the BEVs produced in the short term could easily be designed to be upgradeable to HFCEVs in the future. This could be accomplished by simply replacing the initial battery pack with a fuel cell package designed to be interchangeable (form factor, electrical connections, and specifications) with the previous battery pack. Honda's EV+ BEV and FCX HFCEV share the same vehicle platform, so why can't other vehicles do the same? Why is there no interest in pre-developing this market for hydrogen fuel given the apparent certainty of the hydrogen herd and their spending of our tax dollars on the development and implementation of the hydrogen economy?

Author in his second electric car, a 1976 EVA Metro

# 14

# *Is The Hydrogen Economy Sustainable?*

What constitutes a sustainable solution? Presumably one where the inputs can reasonably be expected to remain available at an attractive price for the foreseeable future, or where the inputs themselves are considered to be sustainable.

In our current context, the term sustainable does not apply to oil and natural gas. It will also exclude coal and uranium if they are to be used in sufficient magnitude to replace the current use of oil and natural gas on the world scale.

The energy inputs that are generally considered to be sustainable (also referred to as renewable) on a planetary scale are geothermal, tidal, solar, wind, hydro, and biomass (the latter four being dependent on solar energy as an input).

There are two primary issues with making the hydrogen economy sustainable.

First, we need to be producing enough primary energy from sustainable sources to meet a significant proportion of the world's energy needs. Clearly, we make good use of renewable energy for passive heating and growing plants, which are essentially beyond our control on a global scale. The only other sustainable source that is currently used to produce a significant amount of energy is hydropower. To date, the degree to which we harness geothermal, tidal, active solar, wind, and biomass energy is trivial. It is hard to imagine that these energy sources will move from their current very small levels of penetration to dominant sources within a human generation or even two (25 or 50 years).

Second, the conversion of any sustainable primary energy source to hydrogen will incur losses. To date, these losses are in the order of 40% to 50% of the primary energy used. So, in effect, the basic cost of using hydrogen is a doubling of our basic primary energy requirements simply to accommodate this supposedly universal storage medium. In a world where our sustainable energy supplies are already inadequate (actually trivial compared to our use of non-renewable sources), doubling our demand is a foolish strategy indeed.

Using any reasonable definition of sustainable, our current and foreseeable production of energy from sustainable sources, and the potential for a pervasive hydrogen economy, which will at least double our energy demand in those sectors where it is used, there is no basis for believing that the hydrogen economy will be sustainable. Hydrogen is produced on a commercial scale today primarily from natural gas and other non-renewable sources (approximately 95% of worldwide hydrogen production). Proposals for the near future are based on electrolysis from nuclear fission or coal-fired generation or reforming of still other non-renewable resources. Taken together, the facts indicate there is no credible basis for expecting hydrogen to be produced predominantly from sustainable sources for the foreseeable future either.

# 15

# *Is The Hydrogen Economy Feasible?*

What constitutes feasible? Technically feasible? Feasible in a practical sense? Economically feasible? All of these?

There is no doubt that it is technically feasible to produce hydrogen, store it, transport it, and feed it to a fuel cell to produce electricity. We have been able to do so for over a century. Thus, the hydrogen energy cycle is technically feasible. Today, and for decades past. The rest is refinements.

However, is it even technically feasible to scale up this energy cycle to a global scale? There are a large number of factors to consider related to hydrogen's characteristics.

Hydrogen is the smallest, lightest atom we know. We won't be able simply to start pumping it through the existing natural gas infrastructure (where that exists), as hydrogen will leak through joints and seals that are resistant to natural gas, and hydrogen embrittles some materials that are impervious to natural gas. In practical terms, a new infrastructure will have to be built.

Hydrogen takes more space to hold a specific amount of useful energy than is the case for oil, gasoline, diesel fuel, methanol, etc. Presuming our energy consumption continues to grow at current rates, converting to hydrogen as the primary energy store will increase the amount of space required for energy storage depots dramatically.

Many credible authors on the subject of energy argue we are reaching a point of scarcity of affordable fossil energy sources (notably oil and natural gas). Significant implementation of the hydrogen economy will dramatically increase the demand for fossil fuels in the short term (as a feed stock and energy source to produce the hydrogen), driving us into a world-wide fossil-fuel energy shortage even sooner than would otherwise be the case.

We live in a world where the North American economy is considered threatened by moderate recent rises in energy prices, notably oil and natural gas. How can anyone believe implementing an energy paradigm that would dramatically raise those prices still further (by significantly increasing the demand for natural gas and energy in general) is feasible in anything like our current circumstances?

There are also many practical considerations related to the actual daily use of hydrogen as a fuel that are cause for concern.

Hydrogen is not a liquid fuel at natural temperatures and pressures at the Earth's surface. For handling purposes, it will either be a cryogenic liquid or a high-pressure gas, both of which present significant hazards. This is considerably different from those fuels to which we are accustomed today; e.g., gasoline, diesel, natural gas, or liquid propane gas.

The extremely cold, liquid storage mode gives rise to additional issues: dealing with evaporation, venting, and static electricity that could give rise to sparks (ignition).

Even those who are trained to handle pressurized hydrogen today are known to have problems with it, which have led to costly, and sometimes hazardous situations.

Clearly, a closer examination of the reality of using hydrogen unveils some unpleasant truths.

# 16

# Hydrogen Fallacies

## 16.1. Hydrogen Cars Will Clean Up the Environment

Hydrogen cars cannot be clean and sustainable unless the hydrogen they use is produced from clean and sustainable sources, not fossil fuels. There is not enough electricity produced today from clean, sustainable sources to electrolyse enough hydrogen for even a modest number of North American vehicles. There will not be sufficient clean, sustainable energy to accomplish this realistically for the foreseeable future.

Therefore, hydrogen must come from non-sustainable sources, such as reforming of natural gas, gasification of coal, or nuclear fission. In this case, hydrogen cars do not provide any appreciable improvement in the quality of the environment, because the fossil fuels are still consumed, and the nuclear waste is still produced.

In one study, Korchinski analysed carbon dioxide greenhouse gas emissions that resulted from the use of hydrogen as a transportation fuel.[1] In this case, natural gas was the primary energy source. Korchinski concluded that "the decline in emissions [was] barely discernable, leading to the conclusion that the reduction in $CO_2$ emissions gained by using hydrogen-powered vehicles is not significant." That's for natural gas, the cleanest of the fossil fuels. In the more likely case that coal is used, carbon dioxide emissions will actually increase.

---

1.   *Fueling America: How Hydrogen Cars Affect the Environment*, Korchinski, William J., November 2004, Reason Foundation.

## 16.2. The Hydrocarbon Economy

Some analysts make the misleading claim that the current fossil fuels energy economy is a hydrogen economy. Fossil fuels, such as oil, coal, and natural gas, are not hydrogen fuels, but hydrocarbon fuels. It is correctly called a hydrocarbon energy economy, not a hydrogen economy. This distinction is crucial to our understanding of the fundamental issues. When hydrocarbons are burned, the carbon is combined with oxygen from the atmosphere to produce carbon monoxide and carbon dioxide. The former is a poison. The latter is a greenhouse gas. Man-made additions to the atmospheric inventory of carbon dioxide resulting from the burning of fossil fuels is linked to observed changes in our climate by most of the world's scientific community.

The existing hydrocarbon energy economy has significant differences from the proposed hydrogen economy. Unlike hydrogen, hydrocarbons exist naturally in significant quantities and take relatively little energy to extract and render into useful forms. Hydrocarbon fuels are energy-dense compared to hydrogen (at ambient temperatures and atmospheric pressure). They are not only an energy source, but also good energy carriers (dense, relatively stable). Hydrocarbon fuels are typically easier to store and transport than hydrogen. There are still appreciable reserves of hydrocarbons on the planet that are fairly easily accessible. There are no hydrogen reserves.

## 16.3. Hydrogen—Distributed Fuel

Despite the musings of hydrogen visionaries, hydrogen will not be a democratic or decentralized fuel. In fact, it will be quite the contrary.

It is unlikely that electrolysers will become common household appliances. Until individual households become independent electricity producers in large numbers, household electrolysers will simply be centrally produced electricity by proxy.

Democratization of energy will only come about when primary energy sources are more evenly distributed and controlled. Hydrogen is an energy carrier, not an energy source, and is irrelevant to the discussion of the democratization of energy production.

Most likely, if the hydrogen economy ever comes to fruition, hydrogen would be produced in large, centralized facilities that are close to primary energy sources. Hydrogen producers could then benefit from economies of scale. They could integrate the operations with either liquefaction or pressurization operations to facilitate the storage and transport of the hydrogen to distribution points or points of consumption. They could also gain from possible synergies of the multiple processes. If production of hydrogen using off-peak power ever becomes economical at the household level, it will certainly be even more lucrative at an industrial scale, especially for large-scale generators of electricity who have the excess electricity in the first place.

We have examined how hydrogen is and will be produced—steam reforming of natural gas and then coal bed methane until those run out, and industrial scale electrolysis thereafter—and how it will be distributed and stored. It should be clear from our examination that this is not going to lead to distributed generation on a large scale.

## *16.4. Home Electrolysis*

Home-based electrolysis is inevitably tied to the concept of hydrogen as a decentralized, distributed, or democratic fuel.

Home electrolysis paints an attractive illusion of energy independence at the household scale. Realistically, however, the household electrolyser is necessary because it is not practical to run hydrogen pipes to every household when the electricity grid is already there, and cheaper to implement and maintain. Unless the household produces the electricity to power the electrolyser, the promise of decentralized energy production is simply a delusion.

Why would consumers spend a significant amount of money for a household appliance that converts electricity to hydrogen? Their other appliances will work just as well on electricity directly, and most will work better (e.g., computers, refrigerators, air conditioners, freezers, microwaves, lighting, televisions, sound systems, sewing machines, etc.). All will work more efficiently by skipping the electrolysis energy conversion step.

Even electric heat will be more efficient than converting electricity to hydrogen, in order to use hydrogen for cooking and heating.

In reality, the only viable application for a household electrolyser will be for fuelling hydrogen powered vehicles, and as discussed elsewhere in this book, this is not nearly as efficient or practical an option as charging a BEV using the same original electricity.

The technology required for the hydrogen economy can be made to work. It works now, and has for decades. Producing hydrogen, storing it, transporting it, and using it are not the important issues. The hydrogen economy is about implementing this technology as a system that is affordable and will continue to work for centuries.

Put another way, just because we can do something, doesn't mean we should do it. For example, most of us can pick up a hammer and use it to smash one or more fingers on our other hand. Most of us recognize that this is not a good idea.

## 16.5. Distributed Generation

Given that using electrolysis-produced hydrogen to generate electricity results in tremendous conversion losses, it should be clear that distributed generation (DG) based on fuel cells is an unrealistic solution.

Better alternatives such as combined heat and power (CHP) systems, based on combined cycle gas turbine technology using ethanol or methanol, already exist. These alternatives can be produced from sustainable sources.

On one hand, hydrogen advocates espouse local production of hydrogen via electrolysis, down to the household level, using power from the grid. On the other hand, other advocates champion hydrogen fuel cells for distributed generation of electricity in order to avoid additional investment in transmission and distribution infrastructure (the grid). These two expected benefits are contradictory.

The energy losses associated with the production of hydrogen, only for it to be transported to another point to turn it into electricity, make this an unattractive option except in very unusual circumstances. For example, where bringing in additional electrical distribution capacity is not physically feasible, and all efficiency options have already been exhausted.

## 16.6. Decarbonization

Some analysts believe that the hydrogen economy is the inevitable end of a progression of decarbonizing hydrocarbon fuels. This illusion is based on coincidence and ignores inconvenient facts.

The decarbonization theory is based on the history of hydrocarbon fuels. Wood was succeeded by coal, which gave way to conventional oil, which is now being supplanted by natural gas. In each case, the number of carbon atoms in the fuel molecules has dropped with each change in this succession. The theory extrapolates that hydrogen, with no carbon atoms, is the logical end of the progression.

This simplistic analysis ignores why this progression of fuels took place. Wood was the first fuel used, because it was ubiquitous, easily harvested and required no special technology. These factors were important in times when transportation was powered by muscles, and an axe was a state-of-the-art technology. Where wood is available, it is still used as a heating fuel, and has not been universally supplanted by lower-carbon fuels.

The major impetus for the transition to coal was the denuding of the forests of Europe, making wood more scarce. Coal is fairly easy to mine and accessible with a pick and shovel (at least it was initially). Coal is more compact than wood, and is easier to transport. It can be used in the same stoves and fire-pits as wood. Even in the steam age, wood was often used where it was available, and coal was used where wood was scarce.

Initially whale oil was used for lighting. As whales became too scarce to supply the market, coal oil was substituted. It could be used in the same lamps as whale oil with minor or no modification. Oil burners for furnaces developed years after the coal oil distribution system was developed, to take advantage of the now abundant, cheap fuel.

Increasing demand for inexpensive oil led to improved drilling and transport technologies (tankers, trucks, pipelines). These technologies also proved amenable to moving natural gas, which was initially considered a hindrance in the search for oil. Natural gas frequently came out the drill holes intended for oil. The fuel with the lower carbon content, natural gas, was not embraced for its higher hydrogen content when it was discovered; it was ignored until oil became less abundant.

The history of the succession of fuels over the past couple of centuries is not explained by the desire for lower carbon content, but rather by the desire for inexpensive energy. As there are no hydrogen wells, the production of hydrogen will always be more expensive than using primary energy directly.

If the decarbonization theory is correct, zero-carbon energy sources of the past such as wind mills, water wheels, and solar heating and lower-carbon energy sources such as biofuels and alcohol, would not have been supplanted by hydrocarbons such as coal and oil in recent centuries.

There is no inevitable progression to hydrogen at the end of the decarbonization progression.

## 16.7. Hydrogen to Electricity or Electricity to Hydrogen?

One faction of the hydrogen herd proclaims that hydrogen is the solution to distributed generation of electricity. They claim that investing in overdue upgrades to the transmission and distribution grid can be avoided by simply putting hydrogen fuel cells in locations where there is an issue with electricity supply.

In contrast, the hydrogen powered vehicle faction exclaims that all we need to do is use electricity to produce hydrogen to power vehicles that won't pollute. Implicit in these statements is the belief that there is either a surplus of electricity or a surplus of hydrogen, or both.

So which is it? Is there so much surplus electricity that we should be using it to power our hydrogen cars, or so much surplus hydrogen that we should be using it to produce electricity? The irony of this push-me, pull-you discussion is that there is not a surplus of either electricity nor hydrogen, and we can't afford the energy conversion losses associated with either strategy implemented on a large scale.

## 16.8. PR Cars

The HFCEV and HICEV vehicles (or public relations [PR] cars) slowly being brought out of the labs of the automakers are just the most recent entries in a procession of hydrogen experiments and pilot projects that go back at least fifty years. None of these projects has been practical enough to go mainstream. The breakthroughs required to render hydrogen practical have not happened. Despite the widespread, taxpayer-funded hype from the hydrogen herd, the break-throughs are still over the horizon. Credible experts who have analyzed what remains to be done, and understand the time it takes to get from development to market and finally to saturation agree that if we had a viable hydrogen energy strategy and complete technology solution today (which we don't have), it would take another fifty years before we would see the real benefits from it being implemented.

## 16.9. Infrastructure

There is a real "chicken and egg" issue that presents a fundamental obstacle to the hydrogen economy. Who is going to take the initial risk to finance it? Will the automakers mass-produce hydrogen-powered vehicles before there is a significant refuelling infrastructure in place? Will vehicle owners invest in home fuelling stations, without the assurance that the vehicles will be affordable and that home will not be the only place they can refuel? Will the energy companies finance a hydrogen-fuelling infrastructure before there is a guarantee that millions of these vehicles will be produced?

So far, it appears that all the champions of the free market economy are content to sit on the sidelines, and let taxpayers pay for this infrastructure. For example, this has been the case with the proposed hydrogen highways in California and British Columbia.

Isn't it odd that the automakers and energy companies that are so adamant about the inevitability and imminence of the hydrogen economy are so reluctant to make basic investments in its infrastructure?

## *16.10. Hydrogen Economy R&D Does No Harm*

When confronted with all the issues that lie ahead for the hydrogen economy, some of its advocates fall back on the defense that spending on the required R&D does no harm. Sadly, this is not the case.

The hyperbole about the hydrogen economy in general, and the Freedom-CAR initiative in particular have killed two important initiatives that could have borne real fruit in a reasonable timeframe: the Partnership for a New Generation of Vehicles and the California Air Resources Board Zero Emissions Vehicle Mandate. The FreedomCAR initiative has more U.S. government funding than all other alternative energy programs combined.

The current focus on the hydrogen economy diverts resources (both financial and intellectual) from seeking alternative solutions. It also diverts the attention of the general population from finding individual measures to improve their situation by lulling them into a sense of complacency. It allows people to believe there is no need to worry; the hydrogen economy is going to single-handedly solve the problems of pollution and energy independence and establishing sources of sustainable energy.

Ironically, this strategy is likely to doom the hydrogen economy. Investment in sustainable energy sources needs to be made now in order to have the energy sources available to produce hydrogen on a mass scale, especially after natural gas becomes scarce.

Remember, hydrogen is not an energy source. If we do not invest in sustainable energy sources on a mass scale now, where will the clean energy required to produce hydrogen come from? Hydrogen must be produced from a clean source for it to be clean itself.

# 17

## *Trust Us—This Time*
## *We Really Mean It*

The premise of this book requires accepting a truly Machiavellian view of North American politics in the early 2000s, especially as it applies to energy and transportation policy. Sadly, there is a rich history that supports taking such a view.

This is not the first time the U.S. government has invested in automotive technologies intended to be radical breakthroughs and environmentally beneficial. Nor is it the first time that North American automakers have promised to deliver a better technology in a few years.

Let's quickly review some similar initiatives by the U.S. government and the carmakers over the past few decades. This list is far from exhaustive.[1]

### *17.1. Electric and Hybrid Vehicle Act*

As a direct result of the first Organization of Petroleum Exporting Countries (OPEC) oil embargo in 1973, the U.S. government realized the vulnerability inherent in its growing dependence on foreign oil, and leaped into action by passing Public Law 94-413[2], the Electric and Hybrid Vehicle Research, Development, and Demonstration Act, into law just three years later in 1976.

This law encouraged industry to develop and refine road vehicles that would dramatically reduce the consumption of imported oil to fuel America's road fleet. It also directed the Administrator to purchase or lease 2,500 electric or electric-hybrid vehicles for demonstration purposes, and purchase or lease 5,000

---

1. Jack Doyle does a good job of covering much of this material in *Taken For A Ride*
2. Bill Summary & Status for the 94th Congress, H.R.8800
   Public Law: 94-413 (9/17/76)
   http://thomas.loc.gov/cgi-bin/bdquery/z?d094:HR08800:@@@L

advanced vehicles for demonstration purposes within six years. (Yes, there were electric-hybrid prototypes around thirty years ago.)

In 1978, the Act was amended to extend the demonstration period to ten years (1986), and the number of vehicles to be acquired to 10,000.

Provision for funding the building of at least 10,000 on-road electric and hybrid vehicles by 1982 was a bold initiative that would have changed America's by-ways for decades to come, and put the U.S. at the forefront of sustainable transportation more than two decades ago.

Naturally, despite a federal law requiring this to happen, there was never a major order for vehicles under this law. Instead, General Electric and Garrett AiResearch each produced two vehicles.

## 17.2. U.S. DOE 2x4 Program

In 1978, continuing under Public Law 94-413, the United States Department of Energy contracted four small companies to build two vehicles each as part of the technology demonstration (Battronic Truck Corporation, Electric Vehicle Associates, Jet Industries, and South Coast Technology). All eight vehicles were pure battery electric, and all used conventional lead-acid batteries. No further vehicles were ever ordered under this initiative.

Combined with the previous four vehicles from General Electric and Garrett AiResearch, this brought the grand total to twelve vehicles acquired under the law, of the 10,000 mandated.

## 17.3. CAFE

The Corporate Average Fuel Economy (CAFE) law in the U.S. requires all automobile manufacturers selling vehicles in the U.S. to obtain a minimum average fuel economy for the entire fleet of vehicles they sell in the course of a calendar year. The intent of these rules was to force the automakers into building more and more fuel-efficient vehicles over time. However, this effort has stalled, and the targets have not moved for several years. The current values for cars (8.6 L/100 km = 27.5 mpg) and light trucks (11.4 L/100 km = 20.7 mpg) came into effect in 1985 and 1996 respectively, although the value for light trucks is almost unchanged since 1987 (11.5 L/100 km = 20.5 mpg). Thus, the initial objectives of improving air and water quality and decreasing dependence on foreign sources of oil have been lost in the Potomac-Detroit two-step.

The North American automakers would have us believe that the CAFE standards set aggressive goals, and they are using all the technological prowess at their disposal to meet them. The 1912 Ford Model T achieved a fuel economy of 6.7L/100 km (35 mpg), 45% better than the mileage obtained by the North American automakers for their light-duty fleets in 2002, the last year for which United States National Highway Traffic Safety Administration (NHTSA) has published figures.

In addition, the CAFE rules have some loopholes you can drive a truck through. Specifically, light trucks: pickup trucks, minivans, and sport utility vehicles (SUVs). Light trucks have a lower set of targets than conventional cars. So, the automakers have elected to make fewer cars (tougher standards) and more light trucks (easier standards), and make those light trucks look more like cars. This has led to several inevitable results. The venerable mainstay of suburbia, the station wagon has all but disappeared, replaced by the minivan and SUV. This was no accident, the light truck minivans were aimed directly at the suburban family market with all their interior fittings and finish and car-like ride. Chrysler, the early leader in minivan sales went so far as to dub their line of offerings the "Magic Wagons" (notably not Magic Vans). For those that disdained the soft, utilitarian lines of the minivans, auto marketers offered the more aggressive-looking, but arguably less utilitarian pickup truck with a station-wagon body, which has since acquired the ironic name sport utility vehicle, a class of vehicles that have almost nothing in common with sports cars, and the utility of which is questionable.

The most daring abuse of this loophole to date is undoubtedly the Chrysler PT Cruiser. Based on the same chassis and drive train as the Chrysler/Dodge Neon compact car, the porky retro-styled PT Cruiser couldn't achieve the CAFE ratings required for cars. So in a stroke of administrative genius, Chrysler executives simply decided to sidestep this legislative inconvenience and classify the PT Cruiser as a light truck instead of a car. Problem solved. In fact, this bit of legerdemain worked so well, Daimler chose to repeat it with the introduction of the 2005 Dodge Magnum hot-rod station wagon.

Continued abuse of this CAFE provision has effectively gutted the original purpose of the rules by making the car a relic and putting many North American drivers into vehicles that qualify as light trucks, but look like cars.

## *17.4. USABC*

Under the auspices of the U.S. government's United States Council for Automotive Research (USCAR), the United States Advanced Battery Consortium (USABC) was launched in 1991. This was a collaborative effort funded by U.S. taxpayers and U.S.-based automobile manufacturers "to pursue research and development of advanced energy systems capable of providing future generations of electric vehicles with significantly increased range and performance".[3]

Unfortunately, despite early apparent successes in the program with some advanced battery technologies (notably NiMH and lithium-ion), the program was soon de-focused into chasing a number of tangents not directly associated with the mission of the program. These diversions included the development of the 42-volt accessory power system standard (an initiative that was essentially still-born), ultracapacitors for load levelling, developing batteries for use in hybrid vehicles (rather than battery-electric vehicles), and funding research in the FreedomCAR fuel-cell development program.

No advanced battery electric vehicles were actually developed as a result of this program. While there were rumours that advanced batteries were developed that met most of the program's stated goals, none actually emerged for commercial use. Any battery advances that were developed are now under wraps in the vaults of the automakers. Despite taxpayer funding of the program, there are no documents published to date regarding advanced battery technologies available to the public from the government-sponsored program.

## *17.5. PNGV*

The Partnership for a New Generation of Vehicles (PNGV) was established in 1993 by the Clinton administration, under USCAR. This program was mandated to use federal government resources in co-operation with the three major U.S. automakers to develop a mass-producible passenger car that would consume the energy equivalent of just 2.9 litres/100 km (achieve 80 mpg gasoline equivalent) within 10 years (by 2004). Well, 2004 has come and gone, and the PNGV cars have not appeared in dealer showrooms. Unfortunately, this is largely because the PNGV program was sacrificed in the name of the hydrogen economy, specifically the FreedomCAR initiative.

---

3.   USCAR USABC Mission and Goals
     http://www.uscar.org/consortia&teams/consortiahomepages/con-usabc.htm

The PNGV program had three primary goals (as stated in the first annual program review in 1994):[4]

1.  Significantly improve national competitiveness in manufacturing.

2.  Implement commercially viable innovation from ongoing research on conventional vehicles.

3.  Develop a vehicle to achieve up to three times the fuel efficiency of today's comparable vehicle (while costing no more to own and drive than today's automobile, adjusted for economics, while meeting the customer's needs for quality, performance, and utility).

Despite being terminated early, this program actually managed to achieve its main goal, the development of a family passenger car that was capable of obtaining 80 miles per U.S. gallon.

## Ford Prodigy

By October 1999, Ford had developed its successor to the P2000 program vehicles, a prototype called the Prodigy.[5]

The Prodigy was a five-seater sedan with a curb weight of 2,387 pounds—considerably lighter than conventional mid-size family cars of the period, which typically weighed about 1,000 pounds more. Aerodynamics were also improved over conventional sedans, with a coefficient of drag of 0.199.

The diesel-electric drive train used a 1.2-liter compression-ignition, direct-injection engine that delivered 74 horsepower (at 4,100 rpm). The engine was accompanied by a 47 horsepower, 3-phase AC motor. An automatic-shift, manual transmission was reported to reduce losses by 20% relative to conventional automatic transmissions.

The Prodigy was able to achieve a fuel economy of 3 L/100 km (78 miles per U.S. gallon) on diesel fuel. This was the energy consumption equivalent of approximately 3.4 L/100km (70 miles per gallon) on gasoline, still short of the PNGV stretch target of 2.95 L/100 km (80 mpg) gasoline equivalent. Plug-in charging of the electrical side of the drive system was not available.

---

4.   *Review of the Research Program of the Partnership for a New Generation of Vehicles,* United States National Research Council, National Academy Press, 1994

5.   Ford Motor Company Press Release, 1999.12.29

Great attention was given to styling of the vehicle exterior and interior, with the expectation that this vehicle could go into production with only minor changes.

## GM Precept

The GM Precept, as displayed in 2000, hit the fuel economy target set for the PNGV program. GM claimed it could achieve 2.6 L/100 km or 90 miles on a U.S. gallon of diesel fuel (the equivalent of 2.95 L/100 km or 80 mpg of gasoline in terms of energy consumption), meeting the PNGV fuel economy target. The Precept prototype was an aerodynamic five-seat, family passenger car, boasting a dual axle drive system that incorporated regenerative braking with an electric drive system (driving the front wheels), coupled with either a diesel engine or a gasoline engine mated to another electric motor (driving the rear wheels).

The diesel engine used was an Isuzu, three-cylinder, 1.3-litre, intercooled and turbocharged unit. It was joined to the rear axle with a 10-kW electric motor from Unique Mobility. The front motor was a 25-kW unit from Panasonic. Both electric motors were 3-phase, 350-volts AC, liquid-cooled, permanent magnet motors. Combined city/highway fuel economy was quoted as 90.4 mpg (or the equivalent of 79.6 mpg of gasoline), or 2.6 L/100 km on diesel, the energy equivalent of 2.95 L/100 km on gasoline.

Two different battery technologies were proposed for a production version of the vehicle: lithium polymer or nickel-metal-hydride. In either case, the battery pack was to have a storage capacity of 3 kWh.

## Dodge ESX3

In the spirit of creating a mass-producible, marketable, and efficient family sedan, Chrysler elected not to start with a clean sheet. Instead, it selected one of its existing vehicle platforms, the Dodge Intrepid, working to invent a new drive train for it, and refining the existing vehicle base.

In February, 2000, the outcome was a five-seat, full-size passenger car that weighed in at just 1,020 kg (2,250 pounds), based on an all-aluminum frame and plastic body panels. Climate controlled seats did the bulk of the cabin heating and cooling, being more efficient than conventional air-based heating and air-conditioning.

The drive train is based on a 15 kW permanent magnet motor/generator, supplemented by a 1.5-litre, three-cylinder, direct-injection diesel engine. A lithium-ion battery pack stores energy from regenerative braking and a generator powered from the diesel engine.

The ESX3 had a fuel economy rating of 3.3 litres/100 km (combined city-highway) or 72 mpg. While the ESX3 did not quite achieve the PNGV fuel consumption target, it was more than just a prototype; it was an example of a mass-producible car, as displayed. DaimlerChrysler executives stated it could be manufactured and sold for a US $7,500 premium over a conventional full-size passenger car.

## PNGV Outcomes

A number of ironies arose from the PNGV program.

1.  When the PNGV program was cancelled, the average fuel economy of the U.S. consumer fleet had actually fallen from 9 L/100 km (26.2 mpg) in 1987 to 9.64 L/100 km (24.4 mpg).[6] This was going in the wrong direction, and a long way from the target 2.95 L/100 km (80 mpg).

2.  Work on hybrids pioneered by the PNGV program with U.S. automakers as partners did not penetrate into their model lines. Japanese automakers excluded from the program actually developed practical hybrid technologies and commercialized them, and were selling them in the U.S. by the time the PNGV program was cancelled.

3.  When the U.S. automakers finally did announce that they would produce hybrids, they were not based on the mid-size cars they developed under PNGV, but rather on light truck models. As of the start of the 2005 model year, none had actually been produced for mass-market sale. Further, when introduced in the 2006 model year in limited numbers, they were based on the Japanese gasoline-electric hybrids, not the U.S. PNGV hybrids.

4.  The mass-market 2.95 L/100 km (80 mpg) car was actually developed and built during the same timeframe as the U.S. PNGV program. However, it was developed by Volkswagen (the Lupo), which was excluded from the PNGV because it is based in Germany. (The Lupo was not introduced in North America.)

Despite the fact that sulphur was considered a major issue with the emissions from the PNGV diesel technologies, biodiesel (which contains no sulphur), was never considered as an option to address the problem.

---

6.  U.S. Department of Commerce, International Trade Administration
    http://www.ita.doc.gov/td/auto/cafe.html

While industry did contribute its agreed portion to the program, most of that was used to fund proprietary product programs, so the results of that work were never made available to the PNGV committee or the public, despite taxpayer funding. The net effect is that the PNGV program became yet another taxpayer subsidy to the North American automakers, in the order of about US $1 billion. To date, it is not apparent that any of the technology developed on the taxpayer's dime under PNGV has made it into conventional vehicles produced by North American automakers. The first hybrid vehicle produced in North America (the Ford Escape, an SUV, not a passenger car) is based on Japanese hybrid technology; not PNGV technology that Ford helped to develop.

In 2001, the last review of the U.S. government's standing committee on the PNGV prior to the dismantling of the program by Presidential directive, stated:[7]

> "The issues addressed by the program are still relevant. The need to reduce the fuel consumption and carbon dioxide emissions of the U.S. automotive fleet is more urgent than ever. Since 1993 there has been a 20 percent increase in the petroleum used in highway transportation, the percentage of U.S. petroleum use derived from imports has risen to 52 percent, and in many parts of the world concerns about the potential for climate change associated with greenhouse gases are even more acute.
>
> "On the other hand, during this time, the demand for sport utility vehicles, vans, and pickup trucks in the United States has drastically increased to the point that they now make up 46 percent of new light-duty vehicle sales. This has increased the importance of reducing the fuel consumption of these vehicles compared to that of the typical family sedan. The EPA Tier 2 emissions standards in all likelihood will increase the fuel consumption of all new cars and threaten to preclude the widespread introduction of the more efficient diesel engine in light-duty vehicles. Lastly, the changed global structure of the industry has made it much more difficult to make sense of U.S. competitiveness statement in Goal 1."

These words still ring true, only more so. The issue around Goal 1 was introduced primarily by the carmakers themselves, by creating international alliances and intertwined ownerships that crossed national borders.

In retrospect, what has the FreedomCAR initiative accomplished that would justify denying the expected benefits of the PNGV program to U.S. taxpayers and vehicle buyers? Certainly, a passenger car that could achieve upwards of 80

---

7.    *Review of the Research Program of the Partnership for a New Generation of Vehicles,* United States National Research Council, National Academy Press, 2001

mpg on diesel today would be better than waiting at least another decade for the first of the hoped-for fleet of hydrogen cars to arrive.

## 17.6. CARB Mandate

In 1990, the California Air Resources Board (CARB) established a mandate for the seven largest car companies selling cars in California. By 1998, 2% of the vehicles they sold were to be zero-emissions, and by 2003 that figure was to rise to 10% of vehicles sold. Contrary to the rhetoric espoused since by both the carmakers and EV advocates, zero-emissions did not necessarily mean electric vehicles, and certainly not just battery-only electric cars. Other technologies that could have been compliant included compressed air, fuel cells, powered roadways, etc.

This was not a case of government trying to pick a winning technology. It was a case of government setting a goal, and telling industry to find a profitable means of achieving it. CARB also did not require the vehicles to be capable of speeds higher than legal speed limits, or have any specific range capabilities or pre-determine how the automakers should market such vehicles.

Rather than focusing on this task, the automakers established a united front against the mandate, and spent many millions of dollars fighting the mandate on all fronts: politically, in the courts, and in a wide-ranging PR campaign that focused on misleading the public about what the automakers had already achieved in the area of BEVs. The automakers could probably have given away neighorhood electric vehicles (NEVs) to meet their 1998 zero-emissions vehicles (ZEVs) targets at a lower cost than they spent fighting the mandate.

Still, despite the obstructions, CARB forced over 1,000 zero-emissions vehicles onto the roads of the state. These were the models provided by the manufacturers under the mandate:

- Ford Ranger EV
- Chrysler EPIC
- GM EV-1
- Nissan Altra, HyperMini
- Toyota E-com
- Honda EV Plus

Informed observers concluded that the automakers went out of their way to make the clean-air vehicles fail, although the technology very nearly succeeded in the marketplace in spite of their efforts. For example, each of the automakers claimed there was no demand for their EVs. However, there was a waiting list for every one of the models made available to the public under the CARB mandate, when each manufacturer pulled their vehicles from the market. This occurred almost immediately after satisfying the minimum requirements set out for them by CARB. The waiting lists existed despite the limited markets where the vehicles were made available (only parts of California, and in some cases, Arizona), the extreme conditions set by the automakers before they would deliver a vehicle to an applicant, and even at the above-market prices established for either sale or lease by the manufacturers themselves.

Despite the availability of advanced NiMH batteries for the EV-1, GM elected to supply the initial vehicles with a type of lead-acid battery with known quality issues, ensuring that performance and reliability would be poor, and tires that had high-energy consumption characteristics when they were cold. This sort of oversight was not atypical amongst the various vehicles provided under the program. If any of these manufacturers supplied conventional vehicles with the equivalent sort of design flaws or quality issues, they would soon be bankrupted by the competition. There were other signs the carmakers were not embracing the spirit of the mandate: installations of chargers were mysteriously delayed, maps of chargers did not reflect locations accurately, chargers were not maintained effectively, a recall program for a trivial part took vehicles out of the hands of those leasing them for months. Most of the vehicle models were never made available for sale to the public; they were either offered only to fleet operators, or only available under lease.

While the Keystone Kops routine on the technology side served its purpose, the automakers had greater success in stifling the clean-air transportation mandate.

By 1996, the automakers succeeded in eliminating all the mandate objectives specific to the original 1998 targets (2% ZEVs). By 1998, the mandate was gutted to the point that most of the ZEV target could be met using non-ZEVs (e.g., hybrids).

By 2003, CARB eliminated the requirement to produce ZEVs at all, in exchange for a promise by the automakers to produce a total of 250 fuel-cell vehicles (FCV) by 2008, combined with a percentage of sales based on fuel-efficient vehicles such as hybrids. None of the 2003 rules would take effect until at least 2005. The requirement for the 250 FCVs does not require that any be made available to the public, offered for sale at all, or even that they be used; only that they be produced. There are already rumours that the requirement for 250 fuel-cell vehicles by 2008 will be delayed, and probably relaxed to include internal combustion engine vehicles fuelled with hydrogen.

Given the history, it seems unlikely that CARB will ever actually force ZEVs onto California's roads in anything beyond public-relations fleet quantities.

## 17.7. The Road Ahead

So, given the track record of the experts when it comes to planning our energy future, doesn't it make you want to take a bit more control over your own energy needs for the future?

Does it strike anyone else as curious that a U.S. administration that is so pro-free-market as to want to privatize Social Security, finds it necessary to subsidize industry in amounts well in excess of US $2,000,000,000.00 just to invest in fundamental R&D for the supposedly inevitable energy currency of the future? If the hydrogen economy is the undeniable way of the future, why aren't the energy companies and automakers investing massive amounts of their shareholders' money to secure a dominant position in the energy infrastructure of the future? Instead, they require the taxpayers to underwrite their tentative forays into this technology.

Could this be tacit proof that there is nothing inevitable about the hydrogen economy, and that governments and major corporations are afraid that if consumers were left to their own devices, they might just find solutions that are more affordable, better suited to their needs, and even ready now?

# 18

## Hydrogen and the Hindenburg

It seems that no book or article on the subject of hydrogen is complete without reference to the *Hindenburg* zeppelin, and specifically its spectacular demise. Who am I to break with convention?

Actually, the *Hindenburg* makes for an interesting analogy to the hydrogen economy. It is another case where reality and perception diverge, due largely to the power of the press. It is also a classic case of where U.S. government actions have invoked the Law of Unintended Consequences in a dramatic fashion.

Let us consider the perception of the *Hindenburg*'s demise. It is a small disaster in human terms (say compared to the sinking of the *Empress of Ireland*, or that less dramatic sinking, the *Titanic*, or an airliner crash).

However, it is seared onto the human psyche because it was the first such tragedy covered live on radio, with visual follow-up coming quickly as newsreel cameras and still photographers were also present. Certain vested interests ensured that the use of hydrogen as the lift medium for the zeppelin got significant mention (as opposed to heavier-than-air aircraft that did not rely on the lighter-than-air gas). Most people who have not studied the topic are under the impression that everyone (or nearly everyone) on-board the *Hindenburg* at the time of its destruction died, and that they were primarily burned in the conflagration of all that hydrogen gas that ignited either spontaneously or due to an errant spark or flame. Some believe the crash occurred at the end of the *Hindenburg*'s maiden voyage across the Atlantic.

The reality is rather different. The *Hindenburg* crossed the Atlantic successfully more than twenty times. (It was the *Titanic* that met its end on its first voyage.) It came after the *Graf Zeppelin*, which logged more than a million miles without serious incident. The *Hindenburg* did not crash. It actually floated to the ground during the course of the fire. The majority of people on-board when it caught fire survived the incident (62 of 97, approximately two-thirds). Of those who did die, most died from injuries resulting from hitting the ground when they jumped from the craft. Only two of those killed died of burns, which was likely caused by the diesel fuel carried on-board to run the engines, or possibly the coating of the skin of the aircraft. It is unlikely that anyone died of burns from the hydrogen, as virtually everyone on board would have been positioned below the hydrogen storage bladders (especially during a docking procedure where focus is on the ground and docking mast). Hydrogen floats upward when released, and hydrogen flames also burn upward.

The initial spark and flame did not originate with the hydrogen, but with the skin of the aircraft. This is a result of the nature of the material in a highly charged electrostatic environment, which was the case on the day in question near Lakehurst, New Jersey as a storm was approaching. This was the opinion of Otto Beyersdorff in 1937[1] and by retired NASA scientist Dr. Addison Bain in the 1990s.[2]

The fire on the ground burned for hours after the hydrogen burned off, fuelled by pools of diesel fuel and wood from the craft's superstructure.

Perhaps the most startling thing for most people is that the root cause of the *Hindenburg*'s demise and the resulting deaths and injuries was the policy of the United States government, right up to the White House.

Even people who have studied the end of the *Hindenburg* are not aware that it (like its sister ships) was designed to use helium as its lift agent, not hydrogen. So, why were the German zeppelins of the 1930s lifted by hydrogen instead of helium? Well, because of the United States government.

Helium is an inert gas that is lighter than air. That means it has a natural inclination to float to the top of the atmosphere. Because it is inert, unlike the reactive hydrogen atom, it is likely to make that trip without bonding with any other substance. Thus, significant supplies of helium near the planet's surface are relatively rare. In the 1930s the major terrestrial source was the natural gas wells near Hugoton, Kansas in the U.S. Not all natural gas wells provide appreciable quantities of helium.

Given the demand for helium as a lift agent for dirigibles and blimps, and the concerns about the military application of the German zeppelins, the U.S. government chose to treat helium as a strategic resource, which they controlled via the Helium Control Act. The key result was that Germany could not acquire any

1.    "The *Hindenburg* Disaster", Chapter 7—What Caused It?
      http://www.awesomestories.com/disasters/hindenburg/hindenburg_ch7.htm
2.    "The *Hindenburg* Disaster, Part 3—What Actually Happened"
      http://americanhistory.about.com/library/weekly/aa042101c.htm

appreciable amounts of helium in the 1930s, and were forced to use hydrogen to lift the zeppelins instead.

While the U.S. government is certainly aware of the benefits expected from the hydrogen economy (see The Allure of Hydrogen chapter), there are two catches to the whole scenario.

1. Hydrogen produced from renewable energy sources would have a negative impact on the primarily U.S. based, multi-national oil companies—major contributors to and supporters of several members of the current administration.

2. The required underlying technology (the mobile hydrogen fuel cell) is not sufficiently mature or robust or affordable for mass use.

Fortunately, for the George W. Bush administration, the announced FreedomCAR initiative doesn't need to be a major issue.

Hydrogen doesn't have to be produced from renewable sources. It can just as easily be produced from steam reforming of natural gas (currently the predominant source), coal gasification, or electrolysis drawing electricity on a massive scale from a resurrected commercial nuclear fission industry or new coal-fired electrical generating stations. The oil and coal industries can support it because hydrogen will effectively be their product, and they get the benefit of wrapping themselves in the energy independence and environmental flags as a bonus. The corporatist PR machine, including government agencies under this Administration plus the mainstream media, which is increasingly controlled by a small group of large multi-national corporations, just won't tell the environmentalists or the public that hydrogen will be a fossil (or nuclear) fuel, once removed.

The North American public is under the impression that the hydrogen economy will lead to less pollution and reduced dependence on foreign oil, which is the opposite of the true situation. Thanks to the spin of the hydrogen economy advocates, and the parroting of that message by a compliant media, few look beyond the hydrogen mirage to see the polluting fossil fuels (natural gas, oil, and coal) used to produce most of it. This green-washing campaign is so complete that almost no one looks at the immense obstacles ahead for a viable hydrogen infrastructure, the overall efficiency of the scenario, or the long time-line before it can be achieved.

To ensure that pesky renewable energy doesn't sneak its way into the equation, the Administration will reduce and divert the already relatively paltry government funding for existing renewable energy and efficiency initiatives. Some will be redirected to support research into fossil fuels (e.g., FutureGen) and nuclear fission reactors under the guise of the shiny, new environmental initiative—the hydrogen energy program. Just enough renewable energy will be supported to serve as window dressing for the real agenda.

In fact, the U.S. Administration used funds from the Department of Energy's renewables program to print their 2001 Energy Plan,[3] which in turn proposed to divert additional funding from this very program.

Even more than the *Hindenburg* accident, George W. Bush's White House and current United States government policy regarding the hydrogen economy will also hurt a lot of people before the story runs its course.

Even if the hydrogen economy could work, it's not going to be in place in a significant way (even according to its proponents) for 25 to 50 years. So what are we going to do for the next generation or two?

The status quo is not a viable option for North America either. Oil cannot remain the predominant energy source for this continent.

---

3.      According to Matthew Terry (2002):

"In order to pay for the cost of printing W's national energy plan (the 2001 Bush / Cheney energy plan), the administration pirated money from the Energy Department's solar and renewable energy and energy conservation budgets. Documents released under court order by the Energy Department revealed that $135,615 was spent from the DOE's solar, renewable and energy conservation budget to produce 10,000 copies of the White House energy plan released last May."

Feeling that those darn hippies at the DOE still had too many funds, $1,317.39 was spent for producing 16 "briefing boards" used by administration officials to illustrate and explain the White House energy plan. The newly released documents also show that $176.40 was taken from the energy conservation program to pay for an Alaska trip by Andrew Lundquist, the White House energy task force's staff director, to promote the energy plan. Just to kick 'em while they are down."

The administration's energy policy, which was printed thanks to environmental protection funds, called for drilling in Alaska's Arctic National Wildlife Refuge at the same time the White House was urging Congress to cut the renewable and energy efficiency research budgets by more than 50 percent", in an article titled "Putting Renewable Resource Money to Good Use"

http://www.beyondmainstream.com/archives/politics/focus/
renewable_resource_money.php

Similar at:
http://www.democrats.org/news/200204030001.html

The planet has taken millions of years to produce the pockets of oil we are consuming now. There is an inevitable sequence. Only after oil is discovered, can it be produced and consumed. You can't pump it if you have not found it.

North American oil discoveries peaked in the 1930s. North American oil production peaked in the 1970s. World oil discoveries peaked in the 1970s. World oil production peaked recently, though it will take a few years before the doubters will come to accept this. It has been over two decades since world oil discoveries (additions to reserves) have kept pace with worldwide oil consumption (depletion of reserves)—and the gap is increasing with time.

Peak Oil is the term that describes the point at which oil production reaches its zenith. The truly important thing about Peak Oil is that after this point, we will produce less oil each year thereafter; no matter how much money and effort we spend on trying to find and produce more. The implications of Peak Oil in an oil-addicted world are definitely significant. If you have not heard the term before, rest assured you will become familiar with it in the near future. Because it is here. I'm not an expert on the subject, but Kenneth Deffeyes is.

On February 11, 2006, Kenneth Deffeyes is quoted as stating:

"…we passed the peak on December 16, 2005."

He is not alone. The subject has some serious implications for all North Americans. Some further reading on the subject:

The Association for the Study of Peak Oil at:

http://www.peakoil.net

*Beyond Oil: The View From Hubbert's Peak* by Kenneth Deffeyes, 2005, ISBN 0-80902-956-1.

"The Hirsch Report" at:

http://www.netl.doe.gov/publications/others/pdf/Oil_Peaking_NETL.pdf

(This document was pulled from the Web site, and restored only after protests. If it disappears again, do a Web search for it, there are several copies mirrored around the Internet now.)

*The Long Emergency* by James Howard Kunstler, 2005, ISBN 0-87113-888-3.

*Powerdown* by Richard Heinberg, 2004, ISBN 0-86751-510-6.

*Twilight in the Desert: The Untold Story of Saudi Arabia's Depleted Oil Fields and What It Means to the World* by Matthew R. Simmons, 2005, ISBN 0-47173-876-X.

Read about it, think about it, reach your own conclusions. It's fascinating stuff, akin to watching a train wreck. Consider the implications. Choose your actions accordingly.

At this point you may be wondering, if Peak Oil is real and important, why you haven't heard about it in the mass media or from your elected leaders. You may have, under its code-name: energy security.

The 1973 and 1978 OPEC petroleum pinches were gentle hints compared to the problems that will result from worldwide demand for oil really exceeding supply, which may already be occurring (exacerbated by the current destruction of Iraq's production and delivery capabilities). Significant remaining reserves of conventional oil appear to lie primarily under countries that are not overly friendly with the U.S. and politically stable, or in less accessible locations (under deep water or in the Arctic).

Oil industry experts love to baffle mere mortals with barrages of data, numbers, and statistics. One measure that critics have tried to use to simplify their lives is the R/P ratio, where R is total proven conventional reserves and P is annual production. The resulting number is the number of years of oil that country has left (assuming the production rate remains stable). There have been heated discussions over these figures. Fortunately, we can ignore all that, because if there is a number that matters, it is not the R/P number, but the R/C number, where C is annual consumption.

For the continental U.S., the R/C number for conventional oil is 5. That's 5 years.[4] That's it, that's all. All the drilling in the Alaska National Wildlife Refuge (ANWR) and offshore could possibly extend that number considerably, by as much as 20%. All the way up to 6 years. In other words, if the U.S. could not import oil, they have about 6 years worth left at current rates of consumption. Which is likely why the experts prefer to talk about the R/P ratio (11 years), or reserves that include unconventional reserves (shale oil, heavy oil, oil sands). But in reality, unconventional oil doesn't really matter, because it takes so much energy to produce it that it is not economically attractive. And really, does an R/P ratio of 11 make you feel much better than an R/C ratio of 6? Or 5?

Fortunately for them, the U.S. has a long history of importing oil, which has kept the R/C and R/P numbers somewhat stable over the past few years. There is another catch to using the R/P or R/C numbers, however. They assume that demand (consumption) for oil is stable year over year.

Demand for oil is not stable. It is rising over time, and at an accelerating rate. Increased demand by China, India, and other emerging national economies is also expected to grow dramatically in the near future. When the experts talk

---

4.    Colorado School of Mines, 2002
      http://hubbert.mines.edu/news/Ivanhoe_02-4.pdf

about the years of proven reserves remaining, they typically base their figures on current consumption figures. If they included realistic, exponentially rising consumption that would reduce their number-of-years-remaining calculations severely.

In North America today, we depend heavily on imports of oil from outside the continent; more than half our current consumption. North Americans are under the impression that we import less oil now than we did in the 1970s, as a response to the OPEC embargoes. In fact, we import more oil now (in real terms and as a percentage of total consumption) than we did in the 1970s. U.S. dependence on foreign oil has increased, not decreased. We appear to be on-track to repeat this dependency for natural gas, if we can install the necessary infrastructure in time.

For natural gas, the U.S. is not in much better shape than is the case for oil. The R/P ratio is 9.5 years, and the R/C ratio is 8 years.[5] This is reflected in consumer pricing already. The price of natural gas in my area rose 136% (more than doubled) from 2001 to 2005 (per the utility company that sells it here). In the past year alone it went up by 26%. At that rate the price doubles in just three years.

With this information, the resurging interest in coal in the U.S. begins to make a certain amount of sense.

Given the state of world politics, it is at least prudent to consider which countries currently hold appreciable reserves of conventional crude oil.

## Saudi Arabia

The largest known reserves are within the control of Saudi Arabia. Ruled by the House of Saud, the current ruling family is increasingly having issues with internal dissent and unrest, including attacks on government targets and foreign nationals by their opponents, notably Islamic fundamentalists. It is worthy of note that most of the actual September 11, 2001 attacks in New York and Washington were carried out mostly by Saudi Arabian nationals. The potential for political upheaval that could destabilize (or conceivably shut off) oil exports from

---

5.   British Petroleum Statistical Review of World Energy, 2004
     http://www.bp.
     com/oiveasstes/bp_internet/globalbp/globalbp_uk_english/publications/
     energy_reviews/STAGING/local_assets/downloads/pdf/
     statistical_review_of_world_energy_full_report_2004.pdf

that nation to North America, even temporarily, has to be considered a real threat.

During the rise in oil prices in 2004 and 2005, Saudi Arabia, the world's swing producer announced on several occasions that they would increase their production. They have not succeeded in doing so.

## Iraq

The second largest estimated reserves are believed to be under the lands of Iraq. However, the recent military occupation of that country by the United States and Great Britain has resulted in marked decreases in oil production, and significant destruction of related assets. Iraq has become a de facto oil products importer in the aftermath of the 2003 invasion. Open warfare between the occupying forces and the indigent population continues. It will take years for a legitimate government to rule Iraq again, especially if the U.S. continues to interfere in attempts to install a regime friendly to their oil addiction. For a clear lesson in this regard, consider the history of Iran in the past sixty years.

## Russia

The third largest reserves of conventional oil are believed to belong to Russia or states within its control. This is the nation (the controlling state in the Soviet Union) labeled "The Evil Empire" by U.S. President Ronald Reagan. Given the history between the U.S. and Russia, one has to question how enthusiastic Russian leaders will be about exporting significant quantities of oil to North America at bargain-basement prices if the opportunity to extract higher prices presents itself. However, the Caspian reserve figures have been revised downward several times now, and the pockets appear to be more fragmented (increasing recovery costs) than originally believed, so Russia simply may not have the reserves for export at all.

## Iran

The fourth largest reserves currently belong to Iran. This is a theocracy ruled by Islamic fundamentalists with a recent history of open hostility toward the United States.

## Other Middle Eastern Nations

Other Middle East states (Qatar, Kuwait, Oman, United Arab Emirates, etc.), taken together also hold significant conventional oil reserves.

## Africa

Some African states have discovered and have begun developing appreciable oil reserves (Nigeria, Chad). These operations have brought little prosperity and much devastation to these nations, to the point that the term "The Resource Curse" was coined to describe the effect.

## Venezuela

The current leader of Venezuela has survived two U.S. sponsored attempts to remove him from power. It's unlikely he's going to put any of the North American Free Trade Agreement (NAFTA) countries on his preferred customer list. In fact, Venezuela is now the major oil supplier to Cuba, and is increasing exports to China. (Incidentally, a study of Cuba may well be instructive on how a country can survive in the post-peak oil era. Not pretty, but it can be done. Of course, it helps if you don't need significant amounts of energy for heating.)

## South China Sea

The actual amount of reserves in the South China Sea area is not well known. This is largely because the area includes a lot of disputed territory. China, which is a significant, and growing, oil importer has designs on this area—and it is not alone. This area is not likely to be a stable source of oil for western nations in the near future. This is the primary cause for the debate over ownership of the Spratly Islands.

## Canada

Canada has already reached its Hubbert's Peak for production of conventional oil. So while it continues to export oil and natural gas to the U.S., it's questionable how long it can continue to do so without impacting domestic needs. Canada still has considerable reserves of oil sands. However, it takes a lot of energy to extract this oil and refine it, possibly more than is embodied in the oil-based products produced. When you consider the value of the water consumed in the process as well (and the energy pumping it), the return may even be negative. As demand for natural gas continues to rise, the oil sands will become increasingly cost-prohibitive as an oil source. In fact, we'll be better off if we use the natural gas directly than use it to produce oil, if maximizing net energy delivered is the goal. If we place any value at all on clean water, we need to keep the groundwater resource as pure as possible for future use.

## *18.1. Scarce Oil Leads to Troubled Waters*

The disruptions don't happen when we start running out of oil. They begin much earlier, when world consumption exceeds production, even briefly. Economic impacts start even earlier, when world consumption exceeds new discoveries. While it is too soon to be absolutely positive, it is my belief we have already reached this point. From the point of view of Hubbert's Peak, it's all downhill from here. Oil prices are going to start rising from now on, as we begin the endgame for consumption of oil reserves on the planet. Ironically, in many parts of the world, we are still flaring off natural gas to get to the remaining oil, rather than actually using this other finite fossil fuel resource.

How dramatic will the price changes be? That I can't predict. However, in 1973, when Saudi Arabia cut production by 10% (at a time when production was exceeding demand worldwide, and Saudi Arabia accounted for a smaller fraction of world production than it does today), the result was a jump in oil prices of 400% (from approximately US $3.00 per barrel to almost US $12.00 per barrel, and in some cases was bid up to US $17.00 per barrel).[6] When demand really exceeds supply, the result will not be short-term upheaval, but real damage to oil-dependent industrial economies. Undoubtedly, the impacts to oil prices will be even more dramatic than what happened in 1973.

In 2004, oil prices moved from less than US $30 per barrel to over US $50 per barrel, an increase of over 70%, at a time when there were no real shortages of oil worldwide, only rumours of possible problems, and a series of hurricanes interrupted production in the Gulf of Mexico for a period of days. In 2006, we have seen oil at US $70 a barrel and higher. If prices are this volatile reacting to rumours and minor supply fluctuations, we can be sure real price increases will be far more dramatic when production begins to fall behind demand. Demand is increasing due to increased U.S. consumption related to larger vehicles in use, and increasing industrialization and economic expansion in China, India, and other nations.

The oil price crisis is approaching, but it is only a harbinger for the real oil availability crisis that will follow. We are doing virtually nothing to prepare for it on a world scale, or within the industrialized countries at a national scale. So when the crisis hits, the impacts will hit full force.

---

6.    *The Seven Sisters*, Anthony Sampson, 1975

Increasingly, folks who know a lot more about the ins and outs of the oil industry are saying the era of peak oil is already upon us, and not decades into the future. The status quo is not a solution; and it's not going to hold.

# 19

## *Prescriptions for Real Solutions*

Now that we have examined the subject to a reasonable extent, you have to wonder:

If the hydrogen economy is the answer,
what the heck is the question?

In retrospect, it appears the proponents of the hydrogen economy left the questions rather vague, probably intentionally. However, if we want to develop real solutions, we have to have a very clear idea of what problems we want to solve.

Here are my best guesses, based on what I have gleaned from the public musings of the hydrogen herd:

- A replacement fuel for diminishing oil reserves
- Independence from the remaining oil reserves controlled by governments that are not supporters of the United States
- Reduced air pollution
- Reduced greenhouse gas emissions

What may be even more curious is whom we have selected as the champions of the new hydrogen economy as a replacement for the old hydrocarbon economy.

First, we have senior levels of government via various research agencies, governmental and at least nominally in the private sector. Given the huge stake that federal, state, and provincial governments have in tax revenues from the sale of fossil-fuel products (gasoline, natural gas, diesel fuel, heating oil, coal, plastics, etc.), and the degree of funding being spent on securing future oil supplies, the alleged degree of interest in replacing fossil fuels is somewhat dubious.

Second, there are the major North American automakers. Here is a set of organizations that have literally bet their companies on the internal combustion engine for approximately a century. But we are expected to believe they are anxious to chuck all that investment to bet on a long shot. That stance goes beyond quixotic; it has to make shareholders sit up and take notice.

Finally, we have the energy, or more realistically, the oil companies. Again, it's hard to fathom the real motivation for this group to abandon their primary product line in favour of something they can't control to nearly the same degree (if we believe the hype about hydrogen coming from renewable sources in order to provide a partial solution to global warming). It makes more sense if we choose to believe the real energy source for the hydrogen economy will be increased consumption of fossil fuels for the foreseeable future.

In short, the nominal leaders of the hydrogen herd are the very actors who have spent the last century getting us into this situation, and have a vested interest in perpetuating the current energy paradigm. It is hard to imagine these parties having a sincere motivation for embracing a hydrogen economy based on sustainable energy sources so much as one second before the last drop of economically-recoverable oil has been squeezed from shale or sand.

Perhaps more worrisome, these actors have developed a mentality and set of textbook answers from their experience. So it is likely they will recycle their past answers as their solutions going forward (e.g., increased use of natural gas, then coal, then nuclear fission using uranium, then breeder reactors through other fissionable elements). It is highly unlikely they will enthusiastically embrace innovative solutions such as increased efficiency and a dramatic shift to sustainable energy sources. In fact, they have actively opposed such measures, even in the past few years.

Putting aside the issues that have already been raised regarding the widespread use of hydrogen, the real problem is that it isn't ready for primetime. Optimistic forecasts suggest that it could start to have a measurable impact by 2030, but more credible estimates put that date at 2050 or later. As we are facing real problems today (2006) in the areas of air quality, climate change, global political stability, etc., do we really have the luxury of waiting three decades or more to implement positive changes?

Perhaps the most unsettling aspect of the current approach to implementing the hydrogen economy in North America is the decidedly socialist approach being taken regarding its financing. In spite of the current U.S. federal administration's declared faith in free enterprise and capitalism, it has found it necessary to provide the lion's share of funding for hydrogen research, development, and implementation. In spite of a philosophy of reducing government expenditures and cutting taxes, this administration still finds it necessary to spend billions of taxpayer dollars to force even the illusion of progress, with only token investment by the private sector. If the U.S. and Canadian governments have so much faith in capitalist economics and free enterprise, why are they using command economy tactics when it comes to the hydrogen economy?

Despite the rhetoric of the energy companies, and companies working on fuel cell research and development, major multi-national companies are decidedly reluctant to invest significant amounts of their own money into developing and refining the necessary technologies. Rather than staking out their claims on various parts of the turf, they are much more interested in having taxpayers fund the work. In California and British Columbia, the "Hydrogen Highways" with their refuelling stations are not being funded by Chevron, Exxon-Mobil, Shell, or the other major transportation fuel retailers, but by the governments (meaning taxpayers). To these multi-nationals, a couple of decades is not a very long time. If they truly believed there were profits to be made in retailing hydrogen by 2010 or 2015, then there would be considerably more evidence of their investing in the required infrastructure using their own money by now. The fact they are leaving this significant undertaking in the hands of various levels of governments speaks volumes.

If the oil economy isn't going to last, and the hydrogen economy isn't a viable solution, at least in the short term, what are we supposed to do? Go back to the technologies of the 1700s or just freeze (or swelter) in the dark? Those options are being postulated by some alarmists, but if I did not have a more positive message to deliver, I would not have bothered to write this book.

I have spent a lot of time examining the proposed hydrogen economy and related topics, and have found them wanting. Fortunately, I feel I am in a position to provide a better answer, due to my own personal experiences, and the wealth of information I have found from others about solutions they have implemented and proved to be viable. However, before I get to the solutions, we need to explore a couple more subjects to establish the context.

## 19.1. The Existing Energy Infrastructure

There is a basic fallacy underpinning our current energy infrastructure—that centralized, homogeneous energy systems (i.e., fossil fuels in very few forms, and a single continental electrical grid without protective cut-outs) are superior to distributed and diverse sources and uses of energy. This is a message promulgated by those that profit from the centralized energy systems (notably privately-owned electrical utilities, petro-fuels industry, coal industry).

This dependence on centralized, homogeneous sources of energy leads to a vulnerability that has been demonstrated on several occasions within the past 40 years.

- In November, 1965, there was the U.S. eastern seaboard blackout.

- In 1973, the United States experienced the first OPEC oil embargo.

- On May 17, 1977, an outage in Miami, Florida affected approximately 1 million people.

- On July 13, 1977, the New York blackout left about 10 million people without power for about a day.

- In 1978, the second OPEC action impacted oil supplies and prices.

- On New Year's Day, 1981 a power outage hit Idaho, Utah, and Wyoming affecting about 1.5 million people for most of a day.

- On December 14, 1994, about 2 million people were affected by a power outage that covered an area from Washington State to Arizona.

- On July 2, 1996, a fallen tree in Idaho caused power outages in fifteen states in the U.S. northwest, and parts of Mexico and Canada for 2 days.

- Just a month later, on August 10, 1996, there was a virtual replay, again causing a U.S. west coast blackout, spreading again to Canada and Mexico, which was attributed to power lines touching trees in Oregon.

- The 1998 Ice Storm (beginning January 4th), obliterated the distribution grid in eastern Ontario, southern Quebec, northern New Brunswick, part of New England, and northern New York state. It affected over 3 million people, leaving them without power in freezing temperatures for a period of hours to weeks in some cases.

- On July 6, 1999, much of New York City was without power again for about a day, as peak demand for air conditioning took power lines beyond the breaking point.

- In 2001, there were rolling blackouts in California as Enron and friends gamed the newly deregulated electricity market in that state.

- The August, 2003 power outage that blacked out most of the northeastern U.S. and Ontario was also blamed on trees, this time in Ohio.[1]

- Also in August, 2003, parts of Arizona were faced with gasoline shortages due to a ruptured pipeline and spill, and subsequent problems.

Admittedly, such major outages have been fairly rare, but they demonstrate the potential for major consequences cascading from minor events (in the case of the August 14, 2003 blackout, lack of maintenance by First Energy meant some untrimmed trees in Ohio, previously identified as a problem, blacked out most of the U.S. northeast and Canada's most populous province.) However, small outages are considerably more common.

In 2002 (before the August 2003 northeast blackout), Ben Carreras of the Oak Ridge National Laboratory was quoted as saying: "from our analysis of data on blackouts in the United States over the past 15 years, we have found that there is on average a blackout every 13 days. We have concluded that the probability of a blackout of a given size decreases slowly as its size increases. In other words, the probability of a large blackout affecting millions of people, lasting eight hours, and representing a large loss of power is smaller than the probability of a small blackout affecting thousands of people for an hour and representing a small loss of power. However, the probability is not as small as expected."[2]

The infrastructure is aging, and responsibility for maintaining it is devolving to fewer organizations due to corporate consolidations. Corporate cost cutting results in less investment in upgrading and maintaining existing facilities. Extreme weather events appear to be increasing in frequency and severity. Taking these factors together, we can expect these outages to increase in frequency, and apparently in magnitude.

In addition, the homogeneous energy infrastructures are usually based on high-quality energy media, even when much lower quality energy sources could be just as effective in specific applications (e.g., space and water heating, which could use pretty much any type of thermal energy). Some examples of lower-quality energy sources are waste wood pieces, solar (heat), heated wastewater, biomass (especially after components of higher value have been extracted, e.g., crop waste from sugar cane, known as bagasse).

Using lower-quality energy sources also has advantages. They are usually cheaper, simply by virtue of the fact they are lower quality energy, and therefore in lower demand. They are more diverse in their availability, and may vary from

1.    Final Report on the August 2003 Blackout in the United States and Canada: Causes and Recommendations, U.S.—Canada Power System Outage Task Force, http://www.nrcan-rncan.gc.ca/media/docs/final/finalrep_e.htm or https://reports.energy.gov/BlackoutFinal-Web.pdf

2.    Oak Ridge National Laboratory Review, Volume 35, http://www.ornl.gov/info/ornlreview/v35_2_02/power_grid.shtml

region to region as to what is easily affordable. Lower quality energy sources are not usually dependent on a fragile infrastructure for distribution.

## 19.2. What Are We Really Trying to Accomplish?

Going back to the State of the Union Address of January 28, 2003, several issues are mentioned that were presumably the justification for the FreedomCAR program: environment, pollution, clean air, and reducing dependence on foreign energy sources.

By now, it should be clear the hydrogen economy as envisaged by the U.S. Administration and the hydrogen herd would not reduce energy consumption, but actually increase it. (If it is not clear, go back to the chapters titled Hydrogen Reality and Hydrogen Efficiency.) Whether it would increase or decrease pollution depends on what one defines as pollution, and what assumptions one makes about the primary sources that will be used to produce hydrogen for use as a fuel. For at least the next two human generations, implementing a significant hydrogen economy will undoubtedly increase the amount of pollution produced on the planet.

# 20

# *Better Ideas*
# *(That Won't Happen)*

We cannot rely upon our governments or major corporations to take effective action on environmental improvement, energy efficiency, or conservation agendas. At senior levels of government, the financial power-base of our elected representatives (and would-be representatives) has been funded by the very entities we would have them control (large energy corporations, e.g., Enron, Halliburton, Exxon-Mobil, Unocal). At lower levels, our representatives simply don't have the resources to compete effectively with these corporations and their armies of lawyers, corporate strategists and public relations people. So, while the levers of power in our supposedly democratic political processes have been usurped by the corporate controllers, it is up to us, ordinary citizens, to exercise our democratic prerogatives and vote with our feet and wallets. You may only get to vote for your elected officials once every 3 or 4 years, but you get to vote through your actions and consumption every day. The bottom line is the corporate ballot box, and a message delivered there will not be ignored. Cast your bottom-line ballots with purpose.

Clearly, the major automakers and oil companies are not going to put the interests of the masses ahead of those of their shareholders and managers, even though the masses are their customers.

There are three classes of actors I have referred to throughout this book: people, corporations, and governments. Each of these are legal persons as well as economic actors. There has also been some work done on the psychology of the corporate person, especially in the past decade.[1] [2]

---

1.    *When Corporations Rule the World,* Korten, David
2.    *The Corporation: The Pathological Pursuit of Profit and Power,* Bakan, Joel

In short, this research has concluded that corporations are sociopathic, even psychopathic, entities, and that it is logical (if not inevitable) that this is the case for lightly constrained (regulated) major corporations.

To my knowledge, no such work has been done with regard to governments, especially senior governments. However, if such an analysis were to be done, I suspect the conclusion would be that unconstrained governments are neither rational, nor sociopathic. They are psychopathic. I'm not prepared to examine why that might be in this book. My personal experience certainly suggests that government departments and agencies inevitably want to expand in terms of resources controlled and power at their disposal. However, I have never seen a government entity willingly concern itself with justifying its existence, or releasing its resources after the need for that organization has subsided.

By way of example, Canadians are still paying national income taxes nominally to pay off the debts incurred by the government of Canada to wage World War I, a war that was concluded more than 85 years ago. This has become a habit. The Canadian federal Liberal government in 1995 imposed a 1.5-cent per litre increase in the excise tax on gasoline sales (from 8.5 to 10 cents per litre) as a temporary deficit reduction measure. The federal government has run budget surpluses since 1998 (to at least 2006). So, the deficit reduction measure has actually been in effect for more years of surpluses now than years of deficits, and remains in effect today (2006).

The most psychopathic act I can conceive of is the conscious killing of multiple people not actually known to the actor. This is an inevitable consequence of modern warfare. The entities that most often start wars are national (senior) governments.

With that as context, let's consider things that our governments could do, if they were truly committed to serving their constituents.

## 20.1. Things Government Could Do

Most of us can figure out there is a problem in the area of energy. Most of us can see that the prices of oil, natural gas, and electricity are rising considerably faster than the rate of inflation, and faster than we can continue to afford in a few years from now. Most of us can figure out that when you have driven to the end of a dead-end road, it's time to change direction. Isn't it odd that our governments can't seem to figure this out?

Perhaps it is because elected governments have given up on anything that will bear political fruit only after the next election. In an era when wielding power is far more important than vision or ideology, when policy is set by campaign contributors and political agendas are set by opinion poll instead of morality or logic, we should not be surprised that our elected leaders have forgotten how to lead. Or perhaps it is because our leaders have given up trying to serve the citizens of their jurisdictions because their real constituency has become lobbyists, voting blocks, and campaign contributors. Whatever the reason, most governments in North America abdicated their responsibility to provide real leadership some decades ago.

We are reaching the end of viable oil, and natural gas will follow in a few years. Even if we weren't reaching these limits, the environmental impacts of our profligate use of fossil fuels in the past couple of centuries clearly signal that we need a better energy source to power the industrialized world.

Spending significant taxpayer money on a grand infrastructure that will use up our remaining oil and natural gas even faster than we are doing now seems a poor short-term strategy, and completely ridiculous in the long term. Yet, that is precisely what our governments are doing today by promoting the hydrogen economy without any plan at all as to how to produce the hydrogen that will be required on a sustainable basis.

Still, if governments wanted to make a difference, they could. No question, making such decisions will have some negative repercussions, but not taking action will have even more severe consequences, if somewhat later.

Hubbert's Peak for production from reserves of conventional oil is real. While the experts continue to claim this won't occur for another decade or four, I believe that examination of the data shows that it has already happened. The changes it will force are inevitable, whether they have already begun to occur in modest forms or will manifest themselves a decade from now or more (most optimistic predictions). Right now, our governments can choose to make a series of planned, manageable changes that will create some dislocation and inconvenience in the short term. Or, they can ignore that reality and leave us to suffer devastating effects that will not be planned or manageable. Experience suggests the latter is more likely.

The following list of helpful things government could do is intended only to be suggestive, and by no means exhaustive.

## Leading By Example

A government committed to sustainable energy could implement the following policy. Beginning in the fiscal year following the announcement of the new policy, that government would increase the amount of energy it consumes from sustainable sources by 10% per year. So, in the first year, 10% of the energy it uses would have to come from sustainable sources; in the second year, 20%; and so on until 100% of the energy it uses (purchasing or producing itself) comes from sustainable sources.

If governments were to do this, with no weasel-out clauses and with a clear definition of what they consider sustainable sources, it is virtually certain that the market would respond to meet that demand. There would likely be some bumps along the road and some initial price instability, but by the end of the decade, the goal would be reached. In particular, I would expect that there would be some shortfalls in the first year or two until the government in question proved its resolve by staying the course in the face of higher-than-market prices (as governments do have a poor credibility track record in such situations). The propensity for new administrations to abrogate the commitments of their predecessors will create another credibility gap to be bridged.

## Higher Prices at the Pump (Carbon Tax)

This one measure alone would do the job. Today, the cost of gasoline or diesel fuel is a trivial fraction of the total cost of operating a car in North America. A carbon tax would implement a clear "polluter-pays" mechanism. The funds from the tax could be used to mitigate environmental damage and implement measures to start reducing our greenhouse gases emissions inventory.

The American Automobile Association (AAA), a pro-automobile organization if ever there was one (it gets its funding from automobile owners), estimates that it cost US $0.562 per mile to operate a car in the U.S. in 2004 on average.[3] Of this, only US $0.065 was related to fuel—or just 11% of the total operating cost. So, increasing the cost of fuel at the pump by 10% will only raise the overall impact of fuel cost to 12% from 11%. A 1% increase in total operating costs is not noticeable in the overall budget, and unlikely to cause any change in behaviour.

In our neighbourhood, I have seen retail gasoline prices change by more than 25% in a 24-hour period, which so dwarfs the concept of a mandated 10% increase as to render it unnoticeable. That 25% jump resulted in grumbling customers, but no perceptible change in consumption patterns. In short, the change in price was so small in real terms as to be ignored by gasoline consumers in terms of the vehicles they drive, or their driving behaviour.

---

3.    http://www.csaa.com/global/articledetail/0,1008010000%257c4512,00.html

Some other driver behaviours have changed. Motorists have become gas-pump gamblers. They check out the posted prices at each gas station they pass, looking to win the litre lottery. If gas prices are relatively low, they fill-up. If they are high, they buy smaller quantities as absolutely necessary. In effect, they all end up playing the short-term fossil-fuel futures market, with their investment limited to how much they can pump into their fuel tank. It has become a game for many, but it simply distracts them from making more fundamental and effective changes.

Further, automotive sales have shown no appreciable decline in demand for gas-guzzlers during this period of rising gasoline prices and increased price volatility. In 2005, there has been an overall slump in demand for new vehicles, but the relative demand for light trucks vs. more fuel-efficient cars has changed little. That means our existing, gas-guzzling fleet may remain on our roads even longer than previous practice (trade-in schedules) would have indicated.

To get an appreciable change in behaviour, i.e., reduction in fuel consumption, a substantial increase in fuel pricing to deliver the required signal to the market would be required, and a widespread belief that the price change is permanent. This change would take a considerable period of time to take effect.

However, given the degree to which North American drivers have come to see cheap gasoline as a birthright, no elected official is likely to have the courage to propose increased gasoline prices if they have any desire to be re-elected. The North American motoring public has demonstrated that they would like a cleaner environment, but they *love* their cars and trucks.

## Effective CAFE Standards

The two-tier standard in the U.S. Corporate Average Fuel Economy (CAFE) standard, copied in Canada, has led to lowered fuel economy instead of improved fuel economy. Vehicles like SUVs, cross-over utility vehicles (CUVs), and the DaimlerChrysler entries like the PT Cruiser, Pacifica, and Magnum have so blurred the distinction between cars and trucks as to render the distinction in the fuel economy rating system meaningless. As a result of the laxer standard for light trucks, the North American automakers have driven the market to vehicles meeting the light truck definition. The consequence is that about half the light-duty vehicles sold in North America today are actually trucks. That shifting of the ratio to more trucks has resulted in the overall fleet fuel economy deteriorating since 1987 to the present day, instead of improving.

Vehicles over 8,500 pounds (3,864 kg) should not be exempted from CAFE rules.

## Clean Up the Smokestacks

The blackout of Ontario and much of the U.S. north-east and mid-west in August 2003 cleared the air on one major issue: how big a role coal-fired electrical generation plays in polluting the air from Ohio to the north and east. Within 48 hours of the grid collapsing, smog had all but disappeared from the affected areas. This happened in spite of the fact that many people were running small generators, and the vehicle population was still capable of full operation. Automotive travel may have been reduced during this period as many workplaces were closed due to the blackout, but it certainly was not eliminated.

Anecdotal reports of reduced respiratory issues at hospital emergency rooms were subsequently confirmed by reduced smog and air pollutant inventories at recording stations, which correlated with the reduced operation of the coal-fired generating stations. Health officials have long been aware of the relation between respiratory distress and diminished air quality. This is why we are advised to avoid heavy exercise on the euphemistically named air quality days.

The suspicions of critics of coal power were substantiated by a study during the period in question by staff at the University of Maryland.[4] The reality was made clear: coal-fired electrical generation is the major source of smog and air pollution in the U.S. northeast, Ontario, Quebec, and eastern Canada.

The solution is simple: clean up the smokestacks.

Governments, as the representatives of the people that suffer the effects of the pollution, need to start a serious program to start retiring the coal-fired plants as quickly as possible, starting with the worst offenders. They then must encourage the remaining plants to start installing technologies to clean up their act. We don't need any advances in these technologies; they are established, proven, and ready now. There are companies in the business of producing scrubbers and other clean-air technologies. All we need is to provide the power plant operators with the incentive to buy them, install them, use them, and maintain them.

My proposal is to provide a four-year time frame (typically the period our senior government politicians are elected to office) for the smokestack industries to clean up their act. Emissions standards and fines would be ramped up in four equal steps, once per year, until the desired standards are attained. Any plant found exceeding the current regulations would be guilty of a violation for each day of operation where they exceeded emissions levels. They would get to choose

---

4.　*The 2003 North American Electrical Blackout: An Accidental Experiment in Atmospheric Chemistry*, Marufu, Taubman, Bloomer, Piety, Doddridge, Stehr and Dickerson, Geophysical Research Letters, May 2004

their most financially attractive course of action (pay the fines, implement the appropriate technology, reduce operations to meet the standards, or shut down the plant unless the market price for power is high enough to cover the cost of operation plus the fines). Fines need to be high enough to get the attention of the corporate executives making the decisions. I suspect something in the order of US $1,000,000 per violation (i.e., per day in violation) might be sufficient.

That is, after the first year of the new regime, fines would be US $250,000 per violation, increasing by US $250,000 per year, until they reached $1,000,000 per violation after year 4. Inspectors will be required to visit each emission point at least twice a day to ensure the monitoring equipment is operating correctly, and not being defeated in any manner. There will be multiple inspectors per plant, and each inspector will be required to visit multiple emission points over the course of a workweek. Areas of responsibility for each inspector will be rotated on a periodic basis, perhaps quarterly.

Fines collected should be deposited to a specific environmental remediation fund, not general government revenues. Initially, this fund could be used to support further R&D on sustainable, clean, generating technologies, and efficiency and conservation measures. While the generating station operators will be entitled to all possible legal avenues of appeal against decisions made against them, they will be required to place the amounts of the fines into escrow accounts at the time of notification of violations, or post equivalent bonds. Interest accrued will go to the same environmental remediation fund. That should remove the incentive for plant operators to stall proceedings in the courts.

## Urban Planning

In North America, we have spent the past fifty years designing our urban infrastructure, and our very society, around cars. If that constitutes planning, it's a terrible plan.

The amount of space we have dedicated to our cars (and light trucks) is staggering; roads, parking, vehicle sales, service, demolition. As much as 30% to 50% of our urban ground space is dedicated to cars.

Some reports estimate that for each car that exists in the urban space, we dedicate a total of six to ten times that amount of space to accommodating that vehicle (a parking space at home, a parking space at work, a parking space at various commercial establishments, and a lot of road space). That's not an efficient use of a scarce resource—urban space.

There are a couple of inevitable consequences to using this space for cars. One is urban sprawl, because we have to put our buildings further apart to make room for the cars and trucks and driveways and streets and roads and parking lots and car dealerships and gas stations and service stations and wrecking yards and so on.

The other is increased reliance on our vehicles, because things are further apart now, which means that walking becomes an inconvenient mode of travel. Walking becomes nigh on impossible as we isolate small sections of communities between high-speed thoroughfares that are impassable to pedestrians and cyclists. This makes the car indispensable for mobility, which increases our energy consumption. Having to use a 2-tonne (4,400 pound) vehicle to transport a 70 kg (155 pound) human and a few kg (pounds) of gear about is hardly efficient.

These two factors feed on each other in an ever-increasing spiral of ground consuming sprawl and energy-sucking practices.

It won't be a light undertaking, but we do need to take back our cities from the cars and trucks so they will once again be fit for human habitation. We can start by shrinking our vehicles to functional (rather than status-defining) sizes.

As occupants of our cities, we have a role to play. We need to use our walkways and those stores and services that are convenient to us as pedestrians to ensure they survive and thrive. We should resist the automatic temptation to jump into our cars to run all our errands when walking or cycling are viable alternatives. Only we ourselves can reduce our degree of dependence on our cars. If each of our multi-car households could find a way to dispense of one of those vehicles (substituting telecommuting, walking, cycling, roller-blading, public transit, car pooling, car sharing programs, rentals, etc.), it would make a huge difference to our air quality, water quality, fitness, and urban spaces.

## Embrace the Electricity Economy

For the areas where the hydrogen economy is supposed to provide the panacea solution, let us consider a more viable alternative: the electricity energy economy. The electricity economy is already a proven quantity. It is robust and extremely reliable. It is already widely deployed, and proven to be economical.

Proven storage options do exist (batteries, hydro reservoirs, pumped storage, etc.), which can be expanded if we choose to do so. Additional stores can be developed (e.g., flywheels, advanced batteries, flow batteries or fuel cells fueled with ethanol). One company doing interesting work in the area of energy storage with flywheels is Flywheel Energy Systems Inc. (http://www.magma.ca/~fesi).

More importantly, the electricity economy provides a base for all the advantages putatively claimed for the hydrogen economy. In fact, it is an intrinsic part of the proposed hydrogen economy; the part that provides most of the advantages desired of the hydrogen energy vision.

Electricity is very efficient in use. Resistance heaters boast nearly 100% efficiency. Electric drive vehicles boast efficiencies in excess of 80% (from on-board electricity to road-wheel).

Electricity produces almost zero emissions in use. There are small electromagnetic fields produced in some applications, and some waste heat in others. However, in general, there are no toxic emissions resulting from the use of electricity in conventional applications (powering motors, electronics, lighting, heating). In the sustainable energy vision of the hydrogen economy, electricity would be the primary source for hydrogen production (electrolysis).

However, the electricity economy produces advantages that are not nearly as probable in the hydrogen economy.

Using electricity directly to power our devices instead of converting it to hydrogen provides a much higher level of overall efficiency. This means that much less primary energy will be required, meaning the conversion to sustainable energy sources can be accomplished sooner and at a lower overall cost to society.

The electricity economy lends itself to distributed generation much more easily than is the case for the hydrogen economy. Significant numbers of people already produce their own electricity, many already using sustainable sources (e.g., wind turbines, photovoltaic panels, low-head hydro). Others produce their own electricity using heat engines coupled with generators, which can be powered from methane, ethanol, biodiesel, or vegetable oil, as well as more conventional fossil fuels.

At a slightly larger scale, community-owned electrical utilities already exist in a number of jurisdictions. This provides residential consumers with the option of producing electricity with economies of scale, while retaining ownership (control) of the production assets and methods. Another means of individuals establishing a connection with electrical generation is to invest in desired (if remote) production assets, such as wind power co-operatives or mid-scale hydropower production, either directly or via mechanisms such as Green Tags. Green Tags are a system where you can ensure that a certain amount of electricity is being supplied to the grid from a sustainable source, e.g., wind power. You pay a fixed amount to the sustainable energy producer to support its operation; in addition to whatever it can make selling the power to the grid. Your financial contribution brings more green power on-line faster.

# Green Tags Ontario   .
## - Clean Air Certificate -

This certifies that

### *Darryl McMahon*

has prevented 1780 kilograms of carbon dioxide entering the atmosphere from Ontario's
coal burning electric generating stations, by purchasing 2 Green Tag(s) from www.greentagsontario.com.

The purchase of a Green Tag produces 1 megawatt hour or more of electricity from renewable wind
energy sources. This Green Tag purchase also prevented 2.2 kilograms of smog causing Nitrogen Oxide,
and 7.2 kilograms of acid rain causing Sulfur Dioxide from entering the atmosphere.
With this act you have made a difference.

**We will all breathe easier.**

**Thank you**

*Authorized Signature, Green Tags Ontario*

Green Tags are a featured product of *The 60/30 Campaign*, a public awareness and action program to reduce greenhouse gas emissions 60 per cent
by the year 2030 in an effort to halt the climate change crisis. Each Green Tag represents an average household reduction of about 5 per cent.

For more information contact

Green Tags Ontario, Box 593, Owen Sound, Ontario N4K 5R4 CANADA
Toll-free: 1-866-545-8414 • Email: info@greentagsontario.com • www.greentagsontario.com

Printed on 100% recycled paper

The major issues facing the electricity economy is a growing demand factor that is not being matched by investment in an adequate, let alone robust, transmission and distribution infrastructure, followed by a lack of sustainable generation facilities. Both of these can be overcome with reasonable funding and a common-sense approach to deploying off-the-shelf solutions. No moon-shots required.

All of this can be accomplished with a fraction of the funding estimated as necessary for implementing the hydrogen economy. The specific targets should be making the transmission and distribution grid more robust, reducing demand on the grid via conservation and efficiency measures, and incentives for the installation of sustainable, distributed (small) generation facilities (e.g., rooftop PV arrays in high insolation areas). In particular, those generation facilities with some degree of storage capacity, which could be used for load smoothing.

## Encourage Energy Diversity

We don't lack for energy and we won't in the future. What we are facing is the impending shortage of the fossil fuels (oil, then natural gas, and eventually even coal) that we have become dependent upon in the past few decades.

However, the sun remains bright in the sky, the winds keep blowing, the rains keep coming, the tides continue to rise and fall, the Earth provides warmth, and trees and other plants still grow. These are the basic energy sources that sustained our evolution from times prehistoric until we discovered coal. They are still available to us today.

Intelligent use of these resources can displace our use of fossil fuels. Increased use of these truly sustainable resources can reduce our reliance on fossil fuels so we can stretch out their availability further into the future for those applications where they are essential.

## Demand Side Management

There was a silver lining to the electricity shortages in the deregulated California market at the turn of the $21^{st}$ century. The good news that came from the rolling blackouts and exorbitant electricity prices in California from 1999 to 2001 was the ability of the ordinary citizens of the state to react to the situation on their own terms. Yes, a few spent thousands of dollars buying generators or photovoltaic systems to provide themselves with more reliable electricity, and got the media coverage. Most, however, simply learned to reduce their consumption of a scarce resource, known in the utility industry as demand side management (DSM). This is the real economics of Adam Smith (essentially the father of the study of economics) and the invisible hand of the free market; the ability of many small economic actors (consumers) to choose their own course of action.

In aggregate, the actions of individuals in California resulted in a drop in electrical consumption at peak demand times of 12 percent from June 2000 to June 2001. That's just one year. Compare that to the time it takes to build generating capacity that would support 12% of demand. The reduction in demand in California was a direct, simple response to the fact that electricity at peak demand periods had become a very expensive commodity. As a consequence of conservation, load-shifting, and efficiency measures, Californians dropped their over all electricity consumption at peak periods by almost 5 gigawatts—the equivalent of ten large conventional (500 megawatt) peaking power plants. Instead of power to the people, which had an ironic ring to it in times of rolling blackouts, we witnessed the power of the people to respond, adapt, and take rational action that was to their benefit.

A motivated public will educate itself and act in its own best interest, for today and the future. In California, they learned that the easiest and fastest way to get control of their electrical bills and keep the lights on was not to start building new power plants, as advised by experts up to and including Vice-President Dick

Cheney. Instead, their best course of action was to reduce their consumption of grid-power through efficiency, innovation, conservation, load-shifting, and alternative sources of energy.

The reality is that government typically relies on experts. Today, our energy experts are almost all drawn from the ranks of major corporations (public or private) that produce big energy from conventional sources. Governments do not hire folks that install insulation or solar heating systems to set policy, but will hire an engineer that has been groomed in the oil industry or the electrical generation industry to make key decisions without hesitation. The experience that is the basis for selecting the experts chosen to make the recommendations skew the process, because their personal beliefs have been shaped by their careers.

Almost without exception, these experts do not believe in demand side management (DSM). Because they have spent their careers developing more and more generating capacity, they have become part of the self-fulfilling prophecy that demand for electricity will continue to rise by 1% to 2% per year forever. As a result, they have ensured that capacity has increased to accommodate that. As a result of that increased capacity, the price of electricity has remained stable (even if it had to be subsidized by our taxes) so that consumers never received a message (price signal) there was any reason to behave differently. The experts then look upon this track record as evidence that demand cannot be managed, that its steady rise is somehow pre-ordained by forces outside their control. They do not recognize that *they* are the force that is driving the steady rise in demand.

Interestingly, when evidence is presented that challenges their belief, the experts will in general dismiss it out of hand, or find cosmetic reasons not to accept it, but will almost never investigate it seriously.

For example, in the city of Woodstock, Ontario, the electrical utility implemented a pre-paid or pay-as-you-go option for its customers in the late 1980s. Under this program, customers were provided with a control unit that showed the instantaneous power consumption inside their homes, and the amount of pre-purchased power they had remaining. When their balance was running low, they would go to a local store, and via a chip-card, buy more electricity, then go home and deposit that purchase into the control unit. If the balance was getting low, the customer had the option of turning things off to make their balance last longer. Customers gained a very real sense of how much power their major appliances (air conditioners, stoves, refrigerators) were drawing, and could modify their behaviour to win the "electrical consumption video game" playing out on their kitchen wall.

This program is still in place, and is now used by approximately 25% of the Woodstock customer base. As subscribers elected to join this program (at an increased cost to them), one result kept coming up, to the surprise of utility officials. *The electrical consumption of these customers dropped by 15 to 20%.*[5] [6] Utility officials looked for explanations, notably theft of power and metering malfunction. Changed consumer behaviour based on having better consumption information simply didn't register as a possible explanation during the initial investigations.

The results in Phoenix, Arizona were almost identical.[7] The experts in the utilities and regulatory bodies involved did not focus on the fact that consumers were able to reduce their consumption of electricity by almost 20% based simply on having better information (demand side management), but spent significant effort looking for other explanations. Even when they were eventually convinced this was evidence that DSM works, no effort to disseminate this information to policy makers, regulators, politicians, or other utilities was made. They certainly did not try to get this information out to the public or their consumers via the media or other means (e.g., their own mail-outs with electrical bills).

This result is evidence that consumers will respond to information that allows them to manage the costs associated with their energy consumption. Receiving a bill from an electrical utility some sixty to ninety days after the consumption has occurred is not useful information. The bill received on a cool day in late September does not help the customer, as she or he would not associate it with air conditioner use on the sweltering days in July. It does not help customers to modify their behaviour by providing a summary several weeks later—the connection does not get made between action and consequence. There is no opportunity to take corrective action; the necessary information comes too late.

Providing good, timely, useful information to electricity consumers is why I am in favour of smart metering in Ontario and elsewhere. However, if we want this initiative to be effective, the units have to be accompanied by a plan for interval (time-of-use) pricing and provision of useful and timely information to the customer. In my view, this information has to include at least all of the following:

---

5.    Presentation by Ken Quesnelle, Woodstock Hydro, Electrical Power Symposium 2004, Ottawa, October 29, 2004

6.    Presentation to Ontario Energy Board
      http://www.oeb.gov.on.ca/documents/sm_woodstock_300804.pdf

7.    Summary of Energy Efficient Programs in Place in Arizona, 2002 (see the entry under "Market Design Initiatives" on page 4)
      http://www.swenergy.org/news/arizona_table.pdf

day-ahead forecasting of electricity pricing by interval, instantaneous display of current consumption, and day-after detailed information as to energy used by interval and derived cost. Without all of these elements as part of the smart metering package, it cannot achieve its potential or the desired result.

There has to be interval pricing, so the customer has the option of shifting her or his electrical demand to off-peak periods to take advantage of cheaper electricity. If electricity costs the consumer the same amount at 3 AM (when there is typically a surplus) as at 6 PM (when demand is typically at its peak), then there is no incentive for consumers to change their behaviour.

Even if the incentive exists, it cannot be used beneficially without appropriate information. If we do not know when the cheap or expensive power is going to occur, we cannot change our behaviour effectively for our benefit, or that of the power generators. This is why the consumer will require day-ahead forecast information. For example, if I know that electricity will cost ten times the average from 3 pm to 7 pm tomorrow, I can take some actions in advance to avoid using power at that time. I can make sure I do not use my air conditioner and my clothes dryer during this time, I can prepare food this evening for a cold dinner the next day, and so on. Similarly, I can choose to do my laundry in the late evening, set the timer on my dishwasher so it will run at 1 AM instead of 9 PM, etc.

Providing the instantaneous display of energy consumption provides consumers with at least two key pieces of information. First, it tells them whether the measures they are implementing are working, right now. So, if the consumption is still high, they can look for additional measures to change the result to the desired outcome. Second, it gives them a very real sense of how much energy and money they are using, and which changes provide them with the maximum benefit. They may not be prepared to stay up until 3 AM to do laundry instead of 9 PM if the difference is only a few cents, but may be prepared to put it off until 9 PM instead of 5 PM if the difference is a couple of dollars and does not constitute a significant inconvenience.

While the more than 2 billion dollars already earmarked for hydrogen energy research and development by North American governments is a drop in the bucket compared to subsidies to the fossil fuel industries, or the wars to control foreign oil, it dwarfs incentives in the areas of sustainable energy. As an incentive to energy producers, 2 billion dollars could launch a successful sustainable energy industry in North America on all fronts—efficiency, conservation, wind, solar, hydro, biofuels, geothermal, etc. As an enhanced Production Tax Credit, it could finance at least 20,000 megawatts of installed capacity of sustainable energy production. That's approximately the equivalent of 40 conventional new natural gas generating stations (500 megawatts each).

By now, it should be clear that our governments are not going to do anything effective. They are responding to the wrong drivers (short term electoral advantage) and listening to the wrong experts. As a result, they will put up some window dressing, and stage some photo-ops (e.g., installing photovoltaic solar panels on some government buildings in a city where cloud cover, smog, and haze will render them largely ineffective, or install a wind turbine where it is highly visible, but poorly sited for capturing wind energy), but we cannot look to government to solve this energy problem for us. We must rely on ourselves to change our behaviour.

## 20.2. Things Governments Actually Do

Curiously, given the far-from-complete litany of things noted above, governments typically choose actions that can only be described as irrational. For example, wide-ranging subsidies to various players that encourage the use of scarce resources or carrying out campaigns aimed at getting people to act irrationally. By and large, people are the only rational actors on the stage of energy production and use.

In general, people do what they are rewarded to do. If I offer to pay you an amount equivalent to your current salary and benefits, now and into the indefinite future, to quit your job and stay home, odds are you will elect to take my offer and stay home. That's because my offer provides you with more free time and lower expenses, while retaining your income. It is rational to do this; you will be better off. This example is hypothetical, I am not making this offer to anyone—I am not a government. This is precisely what some farm programs and social assistance programs do: pay farmers not to plant crops, or pay low-wage earners to stay out of the workforce.

Let me describe a couple of examples for you, after providing you with some relevant context. I live in Canada. We Canadians have the distinction of being the world's largest per capita consumers of energy. Unfortunate, but perfectly rational. You see, we pay (directly) about the least for conventional energy of any nation on the planet. As a result, our citizens very rationally choose to consume more conventional energy, which is cheaper, than environmentally superior, but more expensive (energy-saving) options.

## Electricity Pricing Deregulation

Let's consider the case of electricity pricing in Ontario in the 2002-2004 period. Ontario's government chose to deregulate electricity prices in 2002. Shortly thereafter the inevitable happened (given the history of the situation). Demand exceeded supply, and prices rose—dramatically (California redux). Less than a year before an expected provincial election, this led to a public outcry for relief from the pain in the pocketbooks of residential electricity ratepayers (also known as voters). Within weeks, the government recanted, and imposed a rate cap for residential and small business electricity consumers. But, because the producers were still to be paid at market rates (key to the deregulation regime), the rate cap was subsidized by taxpayers.

The government of the time, and its successor, exhorted residential and small business electricity consumers to reduce their electricity use, as the province was still in a position where it had to import expensive power on a regular basis. To no avail, for Ontario electricity consumers were actually consuming more electricity by August 2004 (adjusted for air conditioning loads related to average temperatures by season) than when the August 2003 blackout occurred, which should have acted as a strident wake-up call. This response was perfectly rational given the subsidized (capped) price of electricity they were paying, and the fact that the government had backed down on market price deregulation previously in the face of an annoyed electorate.

This approach to deregulation discourages investment in the electricity economy because it raises the financial risk, especially for capital-intensive projects. Sustainable energy production tends to be capital intensive, because the fuels are usually free, but the infrastructure to harvest it is more expensive than conventional power plants for the same production capacity. Similarly, a robust and reliable electricity transmission and distribution system is more expensive to install than one with less redundancy and integrity.

## Oppose Clean-Air Initiatives

Off Cape Cod, there is a proposal for the private sector to make a major investment in offshore wind turbines. Instead of encouraging this renewable energy resource, politicians are actively working to block this project with initiatives at both the state and federal levels.

The Clear Skies Initiative of the second Bush administration reduces the effect of the pre-existing Clean Air Act. It actually increases the amount of air pollution produced by electrical power generation in the U.S.

## Discourage Beneficial Industry Change

Another example, this time financed by all Canadian taxpayers, courtesy of the Canadian federal government. In 2002, the Canadian federal government ratified the Kyoto Accord, which aims to reduce greenhouse gas (GHG) emissions. The nominal key to accomplishing this in Canada is reducing our use of hydrocarbon fuels, which produce carbon dioxide, the major greenhouse gas. The actual major producers of greenhouse gases in Canada are industrial sectors e.g., commercial transport, cement production, pulp and paper production, electricity generation, and others. The government did sponsor some voluntary programs to encourage companies to reduce their GHG emissions, and publicized a few of the examples where companies did reduce GHG emissions. But, in general, GHG emissions from industrial sources actually continued to rise in spite of these efforts. And in order to get political buy-in from industry players, the government has effectively guaranteed these companies that the taxpayer will pick up the tab for any penalties (beyond some nominal amounts) that might accrue to these companies for missing targets. So, to the surprise of no one, there has not been a major surge in interest by most Canadian companies to invest in technologies to reduce their GHG emissions.

There have been some showcase examples, e.g., a Canadian pulp and paper operation that has made phenomenal reductions in its GHG emissions over the past decade. However, this outcome was driven by a desire for financial savings that pre-dated Canada's Kyoto commitment. The GHG reductions were just a serendipitous side effect.

## Propaganda Instead of Substance

To counter the lack of effective action on the industrial and commercial fronts on GHG emission reductions, the federal government chose to target another sector: consumers. The One-Tonne Challenge uses electronic and print media (paid for by taxpayers) to exhort Canadian consumers to reduce their direct and indirect GHG emissions by one tonne (of carbon dioxide or the equivalent in other GHGs) per person, or roughly 20%. According to the campaign, this would be accomplished by reducing consumption of electricity (actually only a minor direct GHG contributor in most of Canada, where hydro and nuclear are major primary energy sources for electricity generation) and fossil fuels (e.g., heating oil, natural gas, gasoline, and diesel fuel).

To date, there is every indication that the One-Tonne Challenge did not succeed. It was cancelled by the federal Conservative government in May 2006.

Canadians are not changing their energy consumption behaviour significantly to dramatically reduce their GHG emissions.

The individual consumer is acting rationally—maximizing the benefits she or he obtains in exchange for the resources spent, and ignoring the government's message. It is government that is behaving irrationally, by believing they can overcome this rational response by spending resources (taxpayer money) on the dissemination of illogical propaganda.

If governments really want to change the actions of consumers, they need to change the paradigm so that it is rational for consumers to take up the desired behaviour. This has been the fundamental basis for sin taxes for many years, so it is not that governments don't understand the concept. If we want consumers to reduce their consumption of fossil fuels and electricity (energy) in order to reduce GHGs, then we need to provide a rationale that resonates with them. A package that appeals to their pocketbooks as well as their desire for a better environment in the future is more likely to be successful than a campaign that focuses only on their goodwill. For each desired action (e.g., riding a bike for a 5 kilometre trip instead of driving an SUV), provide the monetary benefit as well as the GHG emissions saving.

One thing that history has taught us is that governments do not solve problems; people solve problems. If you still believe that national governments are effective at much other than spreading tax money to their friends, I suggest you give some serious consideration to the international track record on the drug trade, and in particular the U.S. federal government's history in the war on drugs, the war on poverty, or the war on terror.

It's hard to get the right answer if you are asking the wrong question.

Most of the rest of this book covers things you, as a consumer and individual, can do to make a difference. It may appear that I am asking you to act irrationally. On the contrary, I believe I am forewarning you of inevitable changes to come regarding our energy future, providing you with precious time to make adjustments before they are forced upon us all, so you can be better prepared when the price shocks take place. I hope the advance warning will allow you to make changes in logical (affordable) steps, and not in a panic when conditions are less favourable (and more expensive).

Please rest assured, if you do these things, they will make a difference. To quote an old aphorism; if you think one small individual can't make a difference, you've never tried to sleep in a room with a mosquito. The list is not exhaustive, but it hardly can be in a book of this size. I also believe that taking these actions early will allow you to reap the maximum benefits in the future.

What the hydrogen economy promises is attractive, e.g., sustainable, non-polluting, affordable energy. Even some of the things attributed to the hydrogen economy by its adherents (democratization of energy supplies, distributed generation, increased efficiency, and reduced dependence on foreign energy sources) are desirable. Just because the hydrogen economy is not going to deliver on these promises doesn't mean we should not aspire to them anyway. However, we will have to shift our societal mindsets to achieve these goals.

I am not recommending that we all evacuate the cities, and build off-grid homes and return to a pioneering lifestyle. Modern western society has accomplished a lot of good things (general literacy, hygienic standards, health care, etc.) that are definitely worth retaining. The modern electrical grid is a wonder of engineering and construction, and enriches many lives on a daily basis. Some things need to be fixed, but there is no need to throw out the good with the bad.

If the hydrogen economy can't be delivered to us in an acceptable time frame, and Peak Oil spells the end of the affordable hydrocarbon economy, how are we supposed to achieve these things? That is the focus of the second part of the book. Long story short, there is not a simple, one-size-fits-all answer. The solution will be comprised of doing a lot of small things, and achieving our goals via the cumulative effects of multiple actions. That's why this volume is a book, not a brochure.

I confess that much (but certainly not all) of what I will tell you in the following pages, can be acquired from other sources. However, I hope the gathering of this information into one place, and in this framework may prove of value to you. Perhaps it will motivate you to undertake at least one positive action, however small. That is my objective.

Solving our impending forced withdrawal from our cheap oil habit is not going to be easy. There isn't going to be a silver bullet that is going to fix the oil addiction. It's not going to be painless. However, reasonable preparations can reduce the severity of the adjustments that will be required.

Suppose fossil heating fuels (oil, natural gas) are to be rationed in the future due to shortages (demand continues to rise, reserves in North America are falling). Those that can manage comfortably with the ration because they are prepared (efficient homes, use of renewable heat sources such as solar) will be better off than those who are not prepared. Would you rather take some simple preventative measures in the next few years, or be caught in the crisis when it occurs? I know many people will treat this warning as just one more alarmist proclaiming the sky is going to fall. Believe what you will, but I was one of the minority who forecasted there would be no major calamities resulting from the calendars turning over from 1999.12.31 to 2000.01.01. I can't give you a fixed date when this crisis is going to become critical, because it will occur incrementally, and is probably already in progress.

Will North American culture have to change? Absolutely! But we won't be successful if we try to do everything at once; that will just result in failure on all fronts from lack of focus and resources. As individuals, we need to set priorities for our time and the resources within our control. I believe that means change begins at home.

Other writers (and government agencies) have presented lists of simple things we can do to make things better. Subsequently, others have argued that such efforts are not sufficient, that we need to take dramatic political and social action. They may be right. At this point, I don't think any of us know for sure—the human track record for accurate prognostication is not good. However, we should be in general agreement that reducing our overall energy consumption is a necessary condition for an acceptable future for our species.

The small actions may not be sufficient themselves, but they are necessary as part of the cultural shift to a more rational energy structure. In general, what I am proposing here is a consumer-level version of a no regrets strategy. There is no magic solution to our imminent energy issues waiting in the wings, but even if it turns out there is one, reducing our energy demands on an enlightened, rational, and voluntary basis will do no harm.

You can choose to be more radical if you wish. However, the various things I advocate here are generally within the control of the consumer, and will for the most part reduce your expenses, improve your quality of life, and reduce your exposure to the negative consequences of resource issues to come.

By all means, be politically active and boycott major corporations that you feel are bad corporate citizens. But isn't it hypocritical to criticize others for their profligate or environmentally-damaging behaviour if you are not engaged in better practices to the extent it is reasonable? I'm not advocating martyrdom or hair shirts, just moderation and sensible alternatives. After all, saving resources puts more resources at your disposal to use, as you deem appropriate.

Instead of looking to technology to deliver a miracle solution, or looking to others to solve our problems, we must examine numerous alternatives across the areas of energy production, storage, and use, and which actions we can reasonably implement on our own or in small groups. Those are the topics of the next chapters.

I'm not focusing on global or political solutions, because historically they don't work. Even if they did, and we could be absolutely sure we have the right answers to impose on the rest of the world, we need to get our own house in order first. Why should we expect the developing world to want less than we have?

> *"Never doubt that a small group of thoughtful committed people can change the world: indeed it's the only thing that ever has!"*
>
> —Margaret Meade

# 21

## *Energy Production*

Here's the deal in short form: we are on a fixed energy income, at the planetary level. Our energy is limited by the amount we get from the sun and within the planet. We can save some of that energy (in trees and other biomass, hydro reservoirs, chemical storage, fossil hydrocarbon reserves, etc.) for use at another time and possibly another place. To use a human-scale analogy, we can skimp on lunches to buy a fancy dinner later, but at the end of the month we have to cover the bills from a fixed income.

To continue the personal finances analogy, hydrocarbons are (energy) savings. We have been blowing through the planet's hydrocarbon savings account over the past 3 centuries at an unsustainable and accelerating pace.[1] Nuclear fission is long-term debt. As we use fission to produce energy, we are accumulating a toxic reservoir that we will have to service for millennia to come, which will ultimately consume more energy than was generated for useful purposes.

What we typically consider to be energy production is a trivial, but usually more convenient, subset of that solar and geothermal energy, and typically based on one of the stores mentioned above.

However, the western world has a philosophy of continual economic growth, fueled by the energy we can harness on this planet. We have a fixed supply of energy at the planetary level over time. In contrast, the western corporatist ethic requires growing demand forever. This can't work forever. I'm not saying we can't make further technological advances, and I'm not saying where we are on the global energy supply and demand curves. Only, that if one line is fixed, and

---

1.  Each litre of gasoline originated as approximately 20 tonnes of plant matter. This was buried at a depth of about 2-5 km for some 100 million years. The conditions—warm shallow seas—which created this petroleum do not now exist, so we aren't making the stuff any more!
    Dukes, Jeffrey S., "Burning Buried Sunshine: Human Consumption of Ancient Solar Energy", *Climactic Change 61*, 31-44, 2003

the other line is rising continuously, they are going to cross. And when it comes to something as concrete as energy, consumption can't actually exceed supply in real terms.

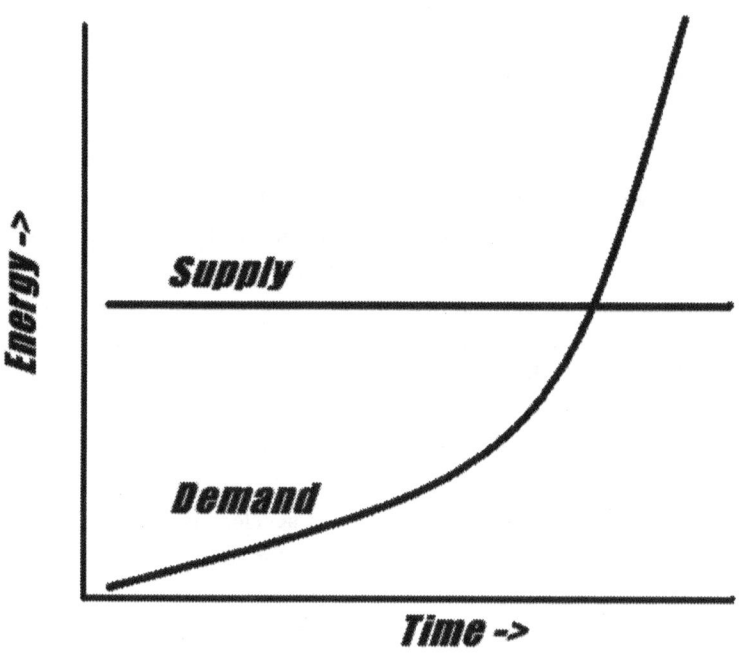

Figure 1

The good news in the short term is we are harnessing a trivial amount of the available energy. The bad news is we are using up the convenient sources faster than we are discovering or producing replacement reserves. In addition, most of the convenient sources are hydrocarbons, and their use is having some dramatic negative consequences on our planet, and by extension, on us. So, because of the negative consequences, and the increasing scarcity of cheap oil and natural gas (and also coal, though this is not as dramatic yet), we need to examine our alternatives. Hopefully, we will find alternatives without the same degree of negative environmental and health effects amongst our options.

At some point, we have to return to living within our energy budget, preferably before we squander our entire savings account. That means using the various sources of sustainable energy available to us, such as geothermal, solar, hydro, wind, biofuels, etc.

If we're smart, we'll elect to start using the sustainable sources extensively in the very near future, so as to stretch the increasingly valuable fossil energy stores as much as possible. Ideally, we would reserve them for those applications where they are absolutely necessary, e.g., we should use geothermal and solar heating to displace the use of natural gas, so that it can be reserved for use as a transport fuel.

Intelligent use of sustainable energy sources means we will not rely on one or two homogenous energy stores, but instead will use all available forms. Regional supplies will vary depending on what is appropriate and available locally. There is no lack of sustainable sources to draw upon. Hydropower is the only source we have exploited to a large extent so far. However, there is still considerable potential in small-scale hydro, including run-of-river and low-head systems. Wind power is the fastest growing sustainable energy source, and it has plenty of room to grow. Photovoltaics continue to make inroads, especially as installed costs continue to fall. Solar thermal systems have a lot of potential, especially for low-tech, residential applications like heating water and buildings. Geothermal energy use will continue to grow, not just on an industrial scale, but also in residential applications in both ground-coupled heat-pump heating and cooling applications, and direct berming (earth-sheltering) of our buildings. The oft-underestimated world of biomass and biofuels will continue to make a growing contribution, especially for heating, but also for transportation fuels.

Often, however, there is a time mismatch between when we want to use the energy and when it is produced from these more desirable sources. This is especially true when it comes to the production and use of electricity, which does not have an implicit storage component, and to a lesser extent, heating. This leads us to the subject of energy storage.

# 22

## *Energy Storage*

We can't always organize our lives so that energy is consumed at the same time and place it is being produced. This introduces the requirement for storing energy, so that we can separate energy use from energy production. This can be as simple as growing a tree, then cutting it down, drying the wood, transporting it to the woodstove, and burning it. The wood provides the energy store, across time and distance.

For the most part, the energy we use is produced elsewhere—be it at an electrical generating station, a petrochemical refinery, or by the sun. Our use of it, at work, at home, and in our vehicles is quite distinct from its production. Clearly energy needs to be stored and transported. However, this leads to a number of questions. What is the best way to store the energy? Who should store it: producers, consumers, or some intermediary third party? Let's examine some of the options.

What makes for a good energy storage option? Presumably, several factors, including initial (capital) cost, on-going (operating) cost, round-trip efficiency, hazards associated with the storage option, space utilized, and potential synergies that accompany the storage option. Some options are better suited to specific energy technologies than others. Pumped hydro reservoirs, for example, make more sense for hydro generating stations than for an oil fired plant, and wood is a good store for producing heat, but not so effective for producing electricity to run a computer.

Historically, the energy producers (electricity, coal, natural gas, transport fuels) have provided the bulk storage for the energy products they have produced. Consumers have provided only very limited storage (e.g., coal bins, heating oil tanks, on-board gasoline and diesel fuel tanks) and relied on frequent refills to keep them going. These technologies and techniques are well established. The success of these strategies provides us with some guidance as to how future energy storage options are likely to work if they are to be successful.

The fact that most transport fuel consumers prefer to buy sufficient fuel to travel 400 to 600 kilometres per refill, rather than keep thousands of litres of fuel at home shows that we are more interested in buying the transportation service than the fuel commodity. Even in times where gasoline prices could fluctuate more than 30% in a week, virtually no consumers chose to put a small fuel depot in their garage to buy fuel when it was cheap in order to protect themselves from higher prices shortly thereafter. Instead, most are content to simply grumble about the high prices and price volatility, which have no apparent relation to crude oil prices or world events. The same day that world crude oil prices fall, local retail gasoline prices rise, and vice versa. While Hurricane Katrina clearly affected refineries and delivery in the southeastern U.S. for a time, the concurrent rise in gasoline prices in Canada is hard to explain given that gasoline stocks are refined and in the distribution channels for months.

In fact, the only consumers who are prepared to invest in any appreciable energy storage facilities of their own are those who are also energy producers. This is the case for those using wood (cords of wood pile), biodiesel home brewers (who may produce 100 litres or so per batch, and might get a couple of batches ahead of their consumption or a bit more if using the brew for home heating as well as transport), and those producing their own electricity off-grid, who typically have a battery bank on the premises.

Why would this mindset change in the future? Why would energy consumers be more interested in storing their own energy when it comes from sustainable sources than is the case today? Frankly, I can see no reason for such a change based on the source of the energy. The only reason I can see for energy consumers to become interested in investing in significant energy storage capacity is where it would save them enough money subsequently to justify the investment, or provide a significant convenience advantage (e.g., sustained power at home in spite of frequent grid outages). For the most part, that seems unlikely.

It is possible that the North American electrical grid will become less stable in the future. This instability will be caused by lack of maintenance and faltering investment in generating capacity coupled with increasing demand. Homeowners may choose to cover themselves against instabilities in the grid by producing or storing their own power. Most are likely to invest in a generator powered by a small gasoline engine and a small reserve of gasoline on-site.

A smaller number might invest in photovoltaics or a small wind generator and a battery bank and will invest in reducing their electrical consumption first.

As fossil fuels become more expensive, homeowners will move away from the fuels to less expensive options, just as they moved from coal to oil, and then from oil to natural gas. As the prices of fuels inevitably rise, people will start to substitute other alternatives, including more efficient home designs, upgraded insulation and weather sealing, other conservation measures, wood and other biofuels, passive and active solar, ground-coupled heating, and seasonal heat storage.

Interestingly, there is no appreciable amount of third-party storage of energy in any area. Even in fields where there are energy brokers, who are neither producers nor consumers, their involvement is primarily in the areas of marketing and finance, not in actual physical storage of energy. There are some third parties (notably trucking companies) that have considerable expertise in the field of transporting fuels, but again, they do not participate in storage of the fuels on a commercial scale.

So, it seems the bulk of energy storage will continue to be provided by the producers as part of the energy service, rather than by the energy consumers.

# 23

## *Energy Use*

North America imported more oil in 2004 than it did in 1974, the year after the first OPEC oil squeeze; both in absolute and percentage terms. Thirty years later, we have not yet learned the lesson of the cost of dependence on an energy source that is not within our control.

## *23.1. Conservation and Efficiency*

I've consciously put conservation and efficiency first, because they are the keys to turning things around, as was proven in California in 2000-2001. In the three R's of recycling—reduce, reuse, recycle—note that reduce comes first.

We have access to some of the richest economies in the world, not because of the rich natural resources at our disposal or because conspicuous consumption is a viable strategy. No, we are still harvesting the benefits of the wisdom of our forebears, who lived by the philosophy of make-do, use-it-up, and wear-it-out. The initial accumulation of wealth that was the springboard to our current material surfeit was not based on consumerism, but on frugality and common sense. We need to return to these values because they are sustainable, better for our overall well being, free us from the shackles of corporatism, and simply because they make sense. The voluntary simplicity movement captures this spirit to some degree, but it is not the message I am after. Instead, my point is that we can maintain our quality of life, and even improve on it, while reducing our energy consumption and shifting from dirtier, finite energy forms to cleaner, sustainable energy forms.

Quality of life (QoL) may be somewhat more difficult to measure than standard of living (SoL), the latter being a gross economic indicator, normally represented as Gross National Product (GNP) per capita. But QoL is a more human concept—it is why people that have accumulated sufficient resources will chuck the "rat race", or why people yearn to retire away from the city back to a simpler, less hectic pace of life. This very reality puts the lie to standard of living (GNP per capita) as a valid measure of how human economies work. Unfortunately, the reliance on standard of living is a driver for our consumer economy. The consumer economy relies on creating artificial dissatisfaction to spur additional consumption with the inevitable consequences of increased consumer consumption, increased energy use (for production and transportation), and increased disposal of surplus goods.

Consider the irony of GNP as the basis for measuring SoL. The building, housing, and maintenance of an Inter-Continental Ballistic Missile (ICBM) with its load of toxic, explosive rocket fuel topped off with a multi-megaton nuclear warhead with a single function—*death and destruction* on a mass scale—counts toward our standard of *living*.

Conversely, a forest provides the essence of life including: filtering our air to produce breathable air; capturing, storing, and filtering our water; moderating the local climate; providing habitat for flora and fauna (food); providing fuel (firewood); and providing us with shelter either directly or to produce building materials. However, this life-giving forest is not counted in our SoL (unless we kill it by razing the trees for a one-time production of construction lumber or pulpwood for making toilet paper).

That's conventional economics as we practice it today. Does that reflect your values as a human being? Which would you rather have in your backyard: an ICBM silo (target for foreign strategic weapons), or a forest?

Despoiling our planet and its future productive capacity to extract short-term riches (consuming natural capital) instead of living on the sustainable bounty of our environment (natural income) is false economy of the highest order. We in the industrialized world need to live within our means. Surprisingly, this is not a bad news message. It isn't hard to do, and it does not require reducing our QoL. It just takes some adjustments, observing and utilizing what works, and a bit of creative thinking.

No matter how efficient we become, we will need to produce some energy. Conservation is just the first step toward a sustainable energy economy. Virtually every credible proponent of sustainable energy use agrees that it is more effective to reduce energy consumption than to install extra generating capacity. Only after we have made the effort to pare down our consumption (demand) should we focus seriously on the supply side of the equation. By reducing our energy consumption, we can reduce the amount of energy required; permitting the amount we produce from sustainable sources to have a greater impact.

It's time to find another viable energy source, or preferably more than one. After all, one-size-fits-all solutions typically fit no one well.

There may not be a "silver bullet" solution for the coming energy crunches, but moving forward with multiple alternatives including improved efficiency, conservation, substitution of sustainable energy sources where appropriate, and incrementing diverse supplies of all practical, sustainable energy sources at the household and community levels (de facto distributed generation) as well as on a large scale should act as a "silver shotgun shell".

## 23.2. Matching Energy Sources to Energy Uses

While hydro resources may be plentiful in a given region, that doesn't mean hydroelectricity should be used for heating. Instead of using a high value energy form like electricity for a low value energy demand like heat, it would be better to use solar thermal energy to the extent possible, supplemented by the electricity to power a ground-coupled heat pump (geothermal) to heat a well-insulated home. The latter approach would match low quality energy sources (solar and geothermal heat) to a low quality energy demand, and reduce the use of the high quality energy form (electricity) so it can be used for other high quality energy demands (e.g., energizing fluorescent lighting, and running electronic devices or electric motors).

# 24

## *Personal Energy Plan*

When I broach the subject of individual responsibility for energy independence or a cleaner environment, the response I receive is almost inevitably one variant or another of the pre-programmed "I can't afford it" response. I have learned this is code for "I don't want to make the effort to think about it".

I have made the effort to think my way through it and have discovered that most people have bought the wrong answer, and are continuing to pay for it—literally!

Reality: You can't afford NOT to start down the road to individual energy independence. It isn't expensive, because it will SAVE you money. However, there is a catch. You will have to think for yourself to make it work.

I can provide some guidelines and examples here, but you will have to figure out how to make them fit into your life. You have to decide which things will work for you, and establish your priorities. I live in a suburban area, in a row house. If you live in a high-rise apartment building or a rural area, things that work for me may not work for you. However, what does work for me or someone else may give you the spark for an idea that will work for you.

I'm not recommending that we all sever our grid connections and produce our own electricity. We don't need to stop using energy. We need to start using it more intelligently.

There are some theories floating around that the amount of energy used per capita is a direct indicator of our QoL. Nonsense!

Your QoL is not improved by using incandescent light bulbs instead of compact fluorescent lights that put out the same amount of light just because they use four times as much energy. Neither is your life enhanced by driving a dump truck as a commuter vehicle instead of a fuel-efficient car, although the dump truck will use a lot more fuel (energy). The amount of energy used does not drive your QoL; the services you obtain from using energy (and other) resources does. Using less energy to achieve the same services will enhance your QoL, because it will leave you with more money or other resources at your disposal.

In North America, we have spent decades subsidizing energy through our taxes. This has provided us with cheap energy and huge profits for the fossil fuels sector.

Proven reserves of easily recoverable oil outside the OPEC nations are dropping, quickly. The western world is already playing politics accordingly.

Proven natural gas reserves within North America are also falling now, and those remaining to be tapped tend to be in the Arctic or offshore. These new reserves require new, massive investments, and additional environment risks to tap into them and transport them to market.

So what do world energy markets and global energy reserves have to do with you? Everything! You are the consumer. The world's energy markets were invented to serve you. You pay for the end product (be it oil, kerosene, heating oil, diesel fuel, gasoline, coal, electricity, wood, alcohol, vegetable oil, wind turbines, photovoltaic panels, insulation, weather-stripping, or whatever). Your purchases drive the energy providers to produce more of whatever energy commodity you are buying, and to develop further upstream resources so that they can keep on supplying and profiting from it.

All the fossil fuels consumed on the planet serve one primary function: satisfying the demands of those willing to pay for them. Consumers. Us.

So when we smugly think about how lightly we're treading on the environment, and all that pollution and greenhouse gases are being produced by "them" (major multi-national corporations, governments, rich people, poor people, suburbanites, etc.), make no mistake: "them" is us.

Whatever your motivation for making beneficial changes to your energy-related behaviour, be it financial, health-related, or altruistic, you need a plan. It doesn't need to be a grand document, or even encompass of all your energy uses. In fact, it is probably better to start small and focused. It's as simple as these four steps:

1. Establish your baseline (current situation)

2. Define an action to improve the current situation

3. Implement the plan

4. Measure the benefit

Then repeat the above (probably in another context). Simpler than building a nuclear reactor, don't you think?

That's my recipe for a positive personal energy plan. Very few of us have the resources (time, finances, physical energy) to take on across-the-board changes. So, start small. Solidify that change. Make another change. Where one change frees up resources, use those resources to invest in the next change. Time is a resource. Use it, which is not the same as wasting it.

It doesn't matter where you reduce your energy consumption or which form of energy. There is energy arbitrage in the overall energy market on the planet. If you reduce your natural gas consumption for home heating or hot water, that leaves more natural gas for electrical generation. If you reduce your electrical consumption, less fossil fuel is used to generate electricity somewhere else on the grid. That fossil fuel can be used for some other application (e.g., transport fuel). So, just pick somewhere to start, and do it.

It will start with an expenditure, of time, personal energy, or money (or more than one of these). Think of it as an investment. Perhaps it is a personal commitment to turning off electricity-consuming items when not in use. Or perhaps switching the most used lights in your home from incandescent to compact fluorescent lights (CFLs). It doesn't have to be big; it has to be achievable.

Suppose you switch four 100-watt incandescent bulbs that are usually lit for five hours per day to equivalent CFLs (e.g., 23-watt units). Your initial outlay would be $20.00 (probably less). Now we wait for the savings to roll in. On your monthly utility bill, your consumption should drop by 45 kWh. If your price for electricity is $0.15 per kWh, that's a savings of $6.75/month, or $81 per year. The CFLs have paid for themselves in 3 months. Plus, there are additional savings due to the longer life of the CFLs. Incandescent bulbs seeing this much service will typically have to be replaced once or twice a year, where the CFL can last 3 to 5 years. If your incandescent bulbs cost $1 each, that's about another $16 saved over the life of the 4 CFLs purchased originally. Suppose the CFLs last 4 years. For your $20 investment, your total savings are $340 (over the four years: [(4 x 81) + 16 = 340]), or **1700%!** Tax-free. Are your retirement savings earning that kind of return?

If that sort of result whets your appetite for more, then you have the option of re-investing your savings in additional ventures that will reduce your energy expenditures. Of course, you don't have to wait for the savings to accrue from your first energy-reduction experiment before launching the next. The sooner you start your energy-reducing projects, the sooner they will start rewarding you with savings.

## Setting Priorities

The priorities you set are entirely up to you. Some people are motivated by financial savings. Others are guided more by reducing their environmental impact. For others it will be something else, perhaps a focus on a specific energy source (negative or positive), or who profits from their consumption of that specific energy source.

On occasion, it may simply be a case of opportunity knocking. For example, at one time I was offered six plastic solar water heating panels free. Building another solar water heater wasn't in my plans at that time, but given the opportunity of a low-cost project and having a place to use it (our cottage), I took the panels with the expectation of using them the following year. Subsequently, I gave some away to someone with a more immediate need, but at least they stayed out of a landfill.

Simple monetary payback may not be the best indicator of real gains because various subsidies and taxes distort the price we pay for various energy sources so that the price bears little relation to the real costs associated with their use. It would undoubtedly be better if we could derive a measure reflecting the quality of the energy being used and the true quantity of the gross energy being consumed. For example, getting heat for our homes (low quality heat) from oil extracted from wells in occupied Iraq is probably a poor return (considering the energy spent on supporting occupying forces and maintaining the oil extraction infrastructure) relative to harvesting sunlight directly to accomplish the same objective. A kWh of electricity is worth more to us than a kWh of heat, as the electricity can be used for many different services (e.g., cooling, lighting, running appliances), and can also be converted into heat with a high degree of efficiency. Unfortunately, it is difficult to establish an index of relative energy values/quality that would not degenerate into an unending debate. In addition, most consumers are already intimately familiar with monetary currencies, but few are conversant, let alone fluent, with energy currencies (e.g., kWh, joules, BTUs, quads). That learning curve alone presents a formidable barrier. In common parlance, when someone says, "just tell me what it costs", or, "what's the bottom line", they are talking about monetary measures and nothing else.

For most of the examples I use here, I will provide the actual energy implications as well as the monetary implications. I will make no attempt to place a realistic value on people's time where that is required to implement a specific change, but rather just note in general terms how much is required when that is relevant.

Most often, I expect the true driver for energy-saving changes, as with most behavioural changes, is a combination of many factors. An array of photovoltaic panels and batteries sufficient to meet all your household electrical needs may be a laudable goal environmentally, but it may not pay off financially or the initial installation cost may exceed your financial resources. Similarly, a wind turbine may seem like a great idea, but perhaps the wind resources at your location don't justify the investment, or perhaps local by-laws prohibit erecting a tower on your property.

The key is to pick an opportunity for changing your energy behaviour that makes sense to you. Then plan for it. Then do it. Bask in the sense of accomplishment. Satisfy yourself that this is a benefit that you can maintain (sustainable progress). Then, when you are ready, find another opportunity and do it again with something else.

We need energy to provide the services we desire. I have given examples of several of the major areas where we use most of our energy, along with the means of reducing our energy expenditures in that area.

These sections are targeted at consumers. Commercial and industrial sectors can also save money via conservation, substitution, and other measures. There are many government, utility, and association programs available to most businesses to help them reduce their energy expenditures. Failing that, there are consultants who can help businesses reduce their energy bills.

For the individual consumer, I have chosen to focus first on energy use, as this is where the biggest gains are available for the smallest investment. Only after addressing personal energy use, do we move on to options for personal energy production, which are typically bigger investments with lower returns.

Those with a significant interest in energy efficiency, conservation, and sustainable energy sources may not find much new content here. What I think is unique is the presentation, from a consumer's point of view, as opposed to that of a product or renewables promoter.

From the perspective of the consumer, the hydrogen economy has been targeted at two primary segments: mobile (transportation) and stationary (distributed generation, building power, power for remote locations). These are also a convenient way to look at the two major areas for personal (consumer) energy use because our major uses of energy are for transportation (our vehicles or using common carriers for longer trips), and our homes (climate control, hot water, appliances and tools, lighting, etc.) We'll consider these in the following chapters.

Many of the changes I cover in the next portion of the book involve physical activity and some do-it-yourself projects. Most of us can afford the extra physical exercise. In fact, the major reason most of us cite for not exercising more is lack of time. So, if I can provide the means for you to exercise regularly without sacrificing an appreciable amount of time, while also reducing your energy costs, it benefits all around.

For those of you without tools or mechanical aptitude, I also have a number of suggestions.

Invest in some good basic hand tools (hammer, screwdrivers, wrenches, pliers, plumber's wrench/water pump pliers). These are sufficient for a large range of household maintenance tasks. For those on a really restricted budget, look for hand-me-downs, at garage/yard/tag/estate sales, try your local freecycle group, or drop some hints regarding gifts near your birthday or other special occasions. If you are not familiar with freecycling, start at http://www.freecycle.org. More later in the book under Waste Management. For basic tools, borrowing should be a last resort.

As for mechanical aptitude, some people are truly gifted. For most of us, it just comes with experience and practice, usually backed with perseverance and determination.

If you are truly convinced you cannot hit a nail or change a light bulb alone, you can always call upon friends who can. Most of us with skills and tools are prepared to help others in need. If you go this route, make sure it is a two-way street. Suppose you are intimidated by the idea of changing your nearly antique showerhead (that rivals a fire-hose for water throughput) for a newer water-saver model, but you are known for your culinary skills. An invitation to dinner is likely a fair trade for the labour you require. Meal-size leftovers are also a good trade, especially for those living alone.

If these approaches don't fit your life or schedule, don't be afraid to hire someone to do the work. Handy people and contractors typically advertise their services, and should not be hard to find.

Every facet of your life where you purchase energy, implicitly or explicitly, is an opportunity for you to reduce your energy consumption, improve your environment, and save money.

# 25

# *Personal Transportation*

North American society has spent over 50 years designing its infrastructure and physical spaces around the car. So it's understandable that many of us think of our cars and trucks when confronted by the words "personal transportation". There are a lot of things we can do with these vehicles to decrease their energy consumption.

Do you want to save money and the environment? For most North Americans, transportation is the number one opportunity you have to do both, and it is simple. *Use internal combustion engine (ICE) vehicles less.*

Most maintenance costs for ICE vehicles are related to the amount they are used. Short trips with lots of stops and at low speeds (typical urban driving) are harder on internal combustion engines than longer, higher-speed trips with few stops (highway driving). Insurance premiums increase with the amount you drive annually. The resale value of your vehicle drops as the odometer reading increases. Oh, and by the way, the use of ICEs is the single largest source of pollution on the planet; air pollution, water pollution, groundwater pollution, thermal pollution, noise pollution, production of greenhouse gases, global warming. The ICE is the big source of all of these problems, and a significant contributor to landfill with all its consumable parts (air filters, oil filters, fuel filters, belts, engine coolant hoses, spark plugs, ignition wires, distributor caps, rotors and points, gaskets, exhaust systems, etc.). The disposal of used automotive engine oil was the single largest source of release of lead into the environment prior to the phasing out of leaded gasoline. Leaded gasoline is still in use in some parts of the world, and in aviation fuel.

Drive your car less. On average, it costs about Cdn $0.53 per kilometer traveled, or US $0.56 per mile driven. More for light trucks. I'm not making up inflated numbers here to get your attention; these are numbers provided by the AAA (2004 figures)[1] and the CAA (2003 figures).[2] These are *pro*-car organizations. That means your 60 kilometre round-trip commute costs you over Cdn

$32.00 *each day*. For those in the U.S., the average daily commute of 40 miles round trip costs you at least US $22.00 per day.

For those concerned with greenhouse gas emissions, each kilometre traveled in the average car produces approximately one-third of a kilogram of carbon dioxide, the major greenhouse gas. For light trucks (including minivans and SUVs), the figure is approximately .45 kg/km. For those of you who prefer Society of Automotive Engineers (SAE) measures, that comes out to 1.2 pounds per mile for cars and 1.6 pounds per mile for light trucks. So, each day you avoid commuting in your SUV or minivan (assuming 60 km/40 miles total daily distance), you can save as much as 27 kg of GHGs (60 pounds).

Of course $CO_2$ isn't the only thing coming out of the tailpipes of our vehicles. There is a whole alphabet soup of other pollutants including: carbon monoxide (CO), nitrogen-oxides ($NO_X$), sulphur-oxides ($SO_X$), volatile organic compounds (VOCs), benzene, toluene, ozone ($O_3$), ammonia ($NH_3$), particulate matter under 10 microns ($PM_{10}$), particulate matter under 2.5 microns ($PM_{2.5}$), polycyclic hydrocarbons, formaldehyde, platinum, etc. In Canada, gasoline may contain up to 1% benzene by volume, and up to 7% toluene is typical.

In addition, the emissions of VOCs, $NO_X$, and $SO_X$ resulting from the extraction, refining, and transport of oil and fuels refined from oil likely exceed the tailpipe emissions for these pollutants.

While catalytic converters and other emissions control devices do reduce the amount of pollutants emitted by vehicles burning gasoline in spark-ignition engines, their effectiveness diminishes with use. Emissions can increase by 25% for each 16,000 km (10,000 miles) traveled, as the emissions control devices deteriorate. After a few years, they are essentially ineffective. Even when new, most operate at reduced efficiency before the engine and exhaust system reach operating temperature. These variables make it difficult to provide a reasonable estimate of how much of each pollutant is produced per litre (or gallon) of gasoline or diesel fuel consumed. However, based on 1997 EPA figures for an average car, here are some figures you can use (again based on a 60 km or 40 mile distance traveled).

1.   California State Automobile Association Web site
     http://www.csaa.com/global/articledetail/0,1008010000%257c4512,00.html
2.   Canadian Automobile Association Web site
     http://www.caa.ca/e/news-issues/btw/2003/btw-03-04-03.shtml

| | | |
|---|---|---|
| Carbon monoxide (CO) | 880 grams | (2 pounds) |
| Hydrocarbons | 116 grams | (4 ounces) |
| Nitrogen oxides (NOx) | 60 grams | (2 ounces) |

Whether you use ethanol blends or regular gasoline, choose the brand that has the lowest sulphur content. No direct savings, but reduced acid rain means your property that is exposed to the weather will last longer. In Canada, according to 1999 figures, it appears that Esso (Imperial Oil / Exxon-Mobil) has the highest sulphur gasoline[3][4][5] and should be avoided. Are you prepared to drive an extra block (no additional out-of-pocket costs) to another service station for the good of the environment?

On cold nights, plug in your block heater so the engine will be warmer for starting. Not only will this make the engine easier to start, but will decrease the time it takes for the engine to come up to temperature—reducing fuel consumption and emissions. Put the block heater on a timer so it only draws electricity for a couple of hours before the vehicle is needed.

Plan your trips to reduce the distance traveled and combine errands. This will reduce the total miles traveled and your fuel consumption costs. Your fossil fuel powered engine works more efficiently and pollutes less when it is operating warm rather than cold.

Check the tire pressure on your vehicles. Under-inflated tires increase fuel consumption and tire wear, reducing life. Over-inflated tires can cause uneven tire wear, and reduce the contact area with the road, and control of the vehicle. Larger tires, e.g., for SUVs and other light trucks, cost considerably more than typical car tires.

Slow down to the posted speed limit. The instantaneous power required to overcome air resistance goes up as the *cube* of speed. It takes about 10% more fuel to cover a given distance at 110 km/h (68 mph) than at 100 km/h (62 mph). Driving at 90 km/h (56 mph) reduces fuel consumption by 20% relative to driving at 110 km/h (68 mph) to travel the same distance. That's why the U.S. intro-

3.    Clean Air Web site
       http://www.cleanair.ca/case_information.html
4.    CBC News Web site
       http://www.cbc.ca/consumers/market/files/cars/lowsulphurgas/table.html
5.    Friends of the Earth Canada Web site
       http://foecanada.org/index.
       php?option=content&task=view&id=15&Itemid=32&limit=1&limitstart=0

duced the national 55 mph speed limit during the OPEC petroleum pinch in the 1970s. Use cruise control, if you have it, for extended highway travel. Observing the posted speed limits won't just save fuel; you can avoid speeding tickets, possibly a collision, higher insurance premiums, and personal injury.

Make sure your vehicle is running properly. Bad ignition components, fouled spark plugs, and dirty air filters will increase fuel consumption and reduce performance.

Car pool. Less gas, less wear and tear, and less parking cost all mean less money spent. Also less air pollution, water pollution, noise pollution, thermal pollution, greenhouse gases, etc. Take note of possible insurance implications from using your vehicle for car-pooling; many insurance companies are improving their policies regarding this practice.

If you commute without your car during the fair-weather seasons, be sure to tell your insurance company. You should probably get a reduced premium for that period.

When it's time to buy your next car, buy the smallest, most fuel-efficient vehicle that is practical for you. If that is a mo-ped, scooter, or motorcycle rather than a car or truck, so much the better. Please don't buy an SUV. Because of their popularity they are outrageously over-priced—the average profit margin on a North American SUV is US $10,000. The recipe is, take a pickup-truck chassis, stick a station-wagon body on it, and then double the price the manufacturing costs would normally justify. If you still have to have one—it's your money, just don't drive it, because it is a prime environment destroyer. SUVs consume a lot of fuel because of their weight, and they are not subject to all the same emission controls or fuel economy rules as regular automobiles, because they are considered light trucks. Consider your real needs and get the appropriate alternative. Need to haul people? Forget the SUV—seating for 2 adults and 3 small children. Need to haul cargo? What real cargo space is there in an SUV? Do you really drive off-road, often? Then get a serious off-roader like an old Willys Overland, International Scout, Dodge PowerWagon, older Jeep or Bronco—not one of the current cosmetic knock-offs. You just have to look at one of the new pretenders with custom paint job, chrome accent trim, designer leather interior, and power play-toys to know these vehicles are not going off-road (intentionally anyway) anytime soon. Due to the increased risk of roll-over due to their high centre of gravity, unintentional off-roading is another matter.

Small engines are actually worse for the environment, as they don't have emission controls. Your gasoline power tools can be replaced with electric tools, such as lawn mowers, or larger devices as required. With a little effort, you can kick the gas-can habit.

Author blowing snow with his electric tractor

## Hybrids

When shopping, consider buying a hybrid, if it is available in the type of vehicle you need. These are a new option from some automakers that use a small electric drive system to help reduce fuel consumption, especially for stop-and-go urban driving. Their gas mileage can be phenomenal, due to the efficient design of the overall vehicle combined with the gains from the hybrid drive-train technology. The Honda Insight is rated at 3 L/100 km or about 70 miles per U.S. gallon. I have friends that have reported getting 90 miles per U.S. gallon with careful driving in the Insight. The first generation Toyota Prius seats 4 adults, and boasts a fuel consumption of just 4 L/100 km (52 miles per U.S. gallon) in *city* driving. I have driven one of these, and can confirm these figures from personal experience. The second-generation Prius is a five-seater.

Note, the Toyota Prius hybrid has enough battery capacity so that it does not require a battery warmer or block heater to ensure starts on cold days. It also provides instant heat courtesy of ceramic electric heaters, so it does not need to wait for the engine to warm up to provide heat for defrosting or warming the cabin.

At least 23 models of hybrid vehicles will be on the North American market by the 2007 model year, based on manufacturer announcements. Unfortunately, many will be very mild hybrids and others will focus on performance (acceleration) rather than efficiency and improved fuel economy. It's frontier time on the hybrid front, so do your research very carefully before making your purchase decision. It appears there will be little on the market to compare to the original Honda Insight (it usually has a waiting list) or either generation of the Toyota Prius.

To date none of the major manufacturers have announced a hybrid based on a renewable liquid fuel (e.g., E85 or biodiesel), so they don't provide a fossil-fuel independent option yet. They remain firmly tethered to the gas pump and imported oil.

Hybrids may not be for everyone. While the hybrid technology reduces fuel consumption markedly in stop-and-go urban driving, it provides no real benefit in highway driving. Depending on the vehicle, other fuel-saving measures may be incorporated (lower co-efficient of drag, reduced frontal area, low rolling resistance tires) to improve highway fuel economy.

## Plug-In Hybrids

Plug-in hybrid electric vehicles (PHEVs) look like a winning technology to me for the near-term (next five to thirty years). So what are they and why haven't you heard of them before? The PHEV is the next evolutionary step along the path from the current hybrid vehicles (e.g., Toyota Prius, Honda Insight, Ford Escape Hybrid) to modern electric vehicles based on advanced batteries. [6] The current hybrids are in reality just electric-assist gasoline vehicles where the electric drive provides a boost to fuel economy, or the more recent trend (e.g., Honda Accord hybrid), higher performance on acceleration. This latter branching will prove an evolutionary dead-end. The PHEV is a hybrid with sufficient battery capacity and electric drive-train capability that it can travel a significant distance in electric-only mode.

Even those familiar with the Prius in North America may be unaware that it has a significant difference from its brethren in Japan and Europe. A feature in the standard second-generation Prius was deleted from the North American model—the stealth button (EV mode in Toyota-speak). This button on the dashboard allows the car to operate as much as possible in electric-only mode; quietly, with zero emissions and no gasoline consumption. Some North Americans found this feature so desirable, they have modified their vehicles to enable

---

6.   The Inevitable Electric Car, Darryl McMahon, Electrifying Times e-zine, 2002
      http://www.electrifyingtimes.com/inevitable_electric_car.html

the feature on their own Prius cars. If that sounds like a good idea to you, then the PHEV will strike you as pure genius.

The idea behind the PHEV is that we can plug in our cars at night (when electricity is in surplus in North America, and therefore cheaper), and then travel a certain distance the next day as a zero-emissions, quiet, oil-independent vehicle. If we need to travel further, then the hybrid capability supplies the energy required for the extended range, albeit now consuming gasoline and producing emissions. How far can it go in electric mode? Well, that will depend on the specific vehicle design, and the amount of battery capacity it carries on-board. Early talk in the field seems to be settling in on an electric range of to 30 to 50 km (20 to 30 miles, sometimes referred to as PHEV-20 or PHEV-30). The higher number is especially interesting, because over 50% of vehicles in North America travel an average of 37 km (23 miles) or less per day. This huge market segment has the potential to become almost emissions-free with the adoption of PHEV-30 vehicles.

Suppose you are a typical North American worker that commutes to work. Let's assume that you live 20 km (12) miles from work. Further, you have to run an errand or two on a typical day, which adds 6 km (4 miles) to your distance. Total distance for the day: 46 km (28 miles). If you could buy an electric car today (lead-acid battery technology), it would meet these needs, but without much margin. The real catch though is that on the weekend, you want to drive to a friend's house, and that's 100 km (60 miles) away. Today's lead-acid battery EV can 't do it. But suppose you had a PHEV-30 vehicle. It could manage your regular daily missions in all-electric mode (no gasoline used, no emissions), but for that weekend mission it can travel the first 50 km (30 miles) as an electric, and then invoke the hybrid drive to get the additional range required. Once at your destination, you can plug the PHEV in to recharge, and then cover half the distance home again in all electric mode, invoking the hybrid capability again to complete the mission. During the course of the week, you will have travelled 430 km (260 miles), but courtesy of the ability of the PHEV, 330 km (200 miles) will have been in electric mode, and just 100 km (60 miles) using fossil fuel. Over three-quarters of your driving will have been emissions-free, independent of foreign oil. Incidentally, the U.S. imports about two-thirds of its oil today. Your PHEV has the potential to make you completely independent of foreign oil—you can manage solely on what is produced at home.

This is a vehicle technology that provides the following benefits:

- The majority of your driving can be zero-emissions

- The majority of your driving can be done free of gasoline

- The majority of your fuel comes from electricity produced at off-peak times (cheaper)

- Most of your refuelling can be done while your car is in the driveway or garage and you are sleeping

- The reduction in gasoline use can make you independent of foreign oil

- When you need more range than the battery can provide, it is available courtesy of the hybrid capability

- When you need fast refuelling, it is available from the existing infrastructure of gasoline stations already in place across the country

- The liquid fuel portion of your energy consumption can be switched over time from gasoline to ethanol blended fuel (gasohol), E85, or other renewable fuels not yet on the market

Too good to be true? The big question that usually comes up is: where will all the electricity come from? Most utilities face a daily load profile where base load (e.g., 3 AM) is about 50% of normal peak demand (e.g., 6 PM). The generation capacity and transmission grid are usually sized to meet the normal peak demand. Finding loads that will occur overnight during low demand times allow the utilities (generators and grid) to utilize their resources more effectively.

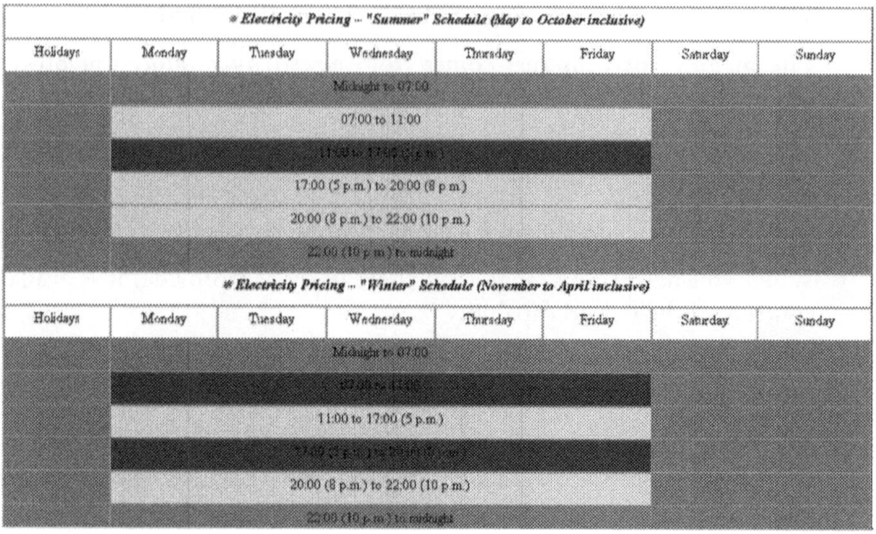

Figure 2

The chart above is the proposed interval pricing structure for electricity in Ontario that started in April 2006 for those with smart meters. The peak periods are the darkest in shading, mid-peak in the lightest shading and the off-peak the middle shading, which is the cheapest—overnight and on weekends. You can see the original at http://www.econogics.com/en/onprical.htm. This figure reflects the daily demand cycle for most of North America.

In 2005, an official of Southern California Edison said that California could accommodate charging of up to sixteen million (16,000,000) PHEVs without adding any additional generating capacity, simply by using the surplus electrical capacity available at night.

Let's take Ontario as an example. According to government figures, daily peak electrical consumption on a typical weekday (not worst case) is approximately 23 gigawatts (between 6 and 7 PM). This is the capacity the system has to be sized for to avoid power outages (actually a bit higher to allow for unplanned outages at a generator, or higher than typical daily peaks). The base load usage can range from 14 gigawatts (minimal heating demand) to 18 gigawatts (cold nights) at approximately 3 AM. The difference is off-peak surplus capacity, roughly 7 gigawatts. The bulk of the daily trough lasts from 10 PM to 6 AM daily, when most people are sleeping. So if we assume just 5 gigawatts is available for this 8-hour period each day, that's 40 gigawatt-hours available for use.

Let's assume a typical PHEV commuter car can get 5 km (3 miles) per kWh, as measured at the charger wall socket. 40 GWh provides us with up to 200,000,000 electric vehicle-kilometres (125,000,000 vehicle-miles) of zero-emissions electric range per day. Suppose our PHEVs travel 35 km (22 miles) per day on average. Our surplus electrical generating capacity could support over 5.7 million such vehicles. With a total population of approximately 10,000,000 people in the province, it's unlikely we have 5.7 million such vehicles. The entire province's population of commuter-mission vehicles could switch to PHEVs overnight without requiring any addition to generating capacity or major infrastructure—just the rolling stock.

If our PHEV commuter car can recharge overnight for seven hours, drawing 7 kWh to supply the recharge for the 35 km travelled the day before, the recharge would take just 1 kW, which can easily be supplied from a regular 120-volt wall socket.

Surplus capacity at night in North America is measured in hundreds or thousands of GWh—many millions of such vehicles could be accommodated with no upgrades to major infrastructure. Yes, some folks might have to install another wall socket to accommodate the charging cord.

Even if the power is generated by coal, the overall impact on the environment is likely to be positive. It's easier to clean up one smokestack than thousands of tailpipes. Historically, tailpipes get dirtier with age (older vehicles fail smog tests), while smokestacks get cleaner over time (retrofit scrubbers etc.).

Interested? Remember Peak Oil? Interested now? Want to know where you can get one? Unfortunately, you can't. Yet. Fortunately, some future-minded folks have already been working on this subject.

Dr. Andrew (Andy) Frank of the University of California at Davis (UC Davis) has been a pioneer in the field. His team has converted vehicles from conventional drive trains to PHEV drive, including a Ford Explorer.

The Prius Plus project at CalCars has modified stock Toyota Prius cars with larger capacity battery packs and on-board chargers to create a PHEV. For more information, check out:

http://www.calcars.org/priusplus.html

EDrive Systems (a co-operative venture of Energy CS and Valence Technology) has done similar work using lithium-ion batteries. An after-market package is said to be in the works. Their Web site is:

http://www.edrivesystems.com/

DaimlerChrysler has several PHEV prototypes based on their Sprinter van in trials around the world. This vehicle is stunning doubters with its 30 km (19 miles) zero-emissions range, and fuel savings of up to 60% in urban driving missions compared to its conventional gasoline counterpart, and up to 50% after the electric range had been used. Each time the van can be plugged in, there is the potential to repeat the 30 km (19 miles) zero emissions range, multiple times per day. Own a fleet of delivery vehicles? Want to save 60% on your future fuel bills? Right now DaimlerChrysler is talking about showroom availability in 2009. You might want to visit your local dealer to let them know you'll be interested in the first units available assuming they are competitively priced.

If you are in the market for something more heavy-duty, check out the EPRI Ford F-550 PHEV trouble truck. Not only does this monster have the ability to recharge at night to provide zero-emissions driving capability, it can also use its hybrid drive system to produce electricity for grid customers while repairing the grid problem. Talk about your win-win product. Check out:

http://www.eei.org/meetings/nonav_2005-05-02-gc/BillWest.pdf

I began construction of my PHEV in 1999. It is not quite finished yet. It is currently licensed and operational as a short-range electric car, getting about 25 km (16 miles) of all electric range using conventional sealed lead-acid batteries. If advanced batteries had been available and affordable, I would have chosen those. The next step is to gather power consumption data, select a diesel generator set, and install it. I expect the generator set will run primarily on biodiesel, thus producing zero net greenhouse gas emissions.

If you want your own PHEV, I can tell you it is not a project for the faint of heart or light-of-wallet; this project has emptied mine. However, if you want to get involved, there are other ways.

The Electric Auto Association (EAA) is a strong supporter of the PHEV initiative. You can join the EAA and support the drive for low and zero emissions vehicles including PHEVs, ABEVs, and the adoption of more vehicles using current technology. The EAA focuses on the folks that are actually putting electric rubber on the road today. EAA's Web site is:

http://www.eaaev.org/

Check out Plug In America, an organization dedicated to promoting EVs and PHEVs at a grass roots, but more political level.

http://www.pluginamerica.com/

Plug-In Partners has kicked off its national (U.S.) campaign (September 2005) to bring communities together to support and acquire PHEVs. Pioneered by Austin Energy and the City of Austin, TX, this organization is focused on municipalities, businesses, non-profits and others that operate fleets as the first point of adoption for PHEVs. Their Web site is:

http://www.pluginpartners.org/

Check out CalCars, a superb source of information about PHEVs. CalCars are the developer of the Prius+ mentioned earlier.

http://www.calcars.org/

If you want to participate in the commercial production of PHEVs, you need to look into the Plug-in Hybrid Development Consortium.

http://www.hybridconsortium.org

There is even a company (HyMotion) building an after-market kit to modify a stock Toyota Prius or Ford Escape Hybrid to a PHEV (there may be other companies, but this is the first of which I am aware).

http://www.hymotion.com/

## Alternate Fuel Vehicles

Beyond hybrids, there are several classes of vehicles that use fuels other than gasoline or diesel oil.

## Ethanol

The most common of these in the car category is likely to be ethanol. While it is not well known, the automakers have received a huge benefit under the CAFE fuel economy rules for producing dual-fuel or flexible-fuel vehicles. Most of these are vehicles capable of running on gasoline, ethanol, or any combination of the two. There have been literally millions of these cars produced in North America in the past decade. However, due to the almost zero availability of E100 or E85 (pure or 85% ethanol) fuels at the retail level, even the owners of these vehicles are often unaware that their vehicles have this ability. E100 has starting issues at temperatures below 13°C, the flash point of ethanol. The 15% gasoline in E85 overcomes this problem.

When purchasing a new vehicle, ask the dealer about ethanol or flex-fuel options. All major North American vehicle manufacturers offer flex-fuel as at least an option on some models. In some cases, flex-fuel vehicles have been delivered without the knowledge of the purchaser (as the manufacturer had to produce and sell a certain number to meet CAFE targets).

When looking for a used car or light truck, consult the owner's manual to see whether it is a flex-fuel vehicle. The current owner may not even know.

You don't have to wait until you have a flex-fuel, E100, or E85 vehicle to use ethanol. Use ethanol blend fuels instead of normal gasoline. Increased demand for ethanol should result in economies of scale in production, thus eventually lowering the cost below that of normal gasoline on a per unit basis. At many retailers, ethanol blends now retail for the same price as regular gasoline. As for the environment, ethanol blend fuels produce less noxious emissions than normal gasoline. Ethanol is an inherent gas-line anti-freeze, and a fuel system cleaner, without the extra fuss and cost. Virtually every gasoline engine on the road today can use conventional ethanol blends up to 10% ethanol by volume (E10) without issues or modification.

## Biodiesel

In the pick-up truck segment, the most likely alternative fuel is biodiesel. There are a limited number of passenger cars available with diesel engines in North America, notably those from German manufacturers. This is unfortunate as diesel vehicles typically get much better fuel economy per litre than gasoline vehicles, on the order of a 30% improvement for vehicles of similar size and weight. Use of biodiesel reduces most noxious emissions considerably relative to regular diesel (e.g., $SO_X$).

Most vehicles designed to run on diesel fuel can use biodiesel blends in any ratio without issues or modifications required. Switching to 100% or high biodiesel ratio blends can present three potential issues:

1.  Biodiesel is a potent fuel system cleaner. If there is a lot of accumulated junk in your fuel system now, the pure biodiesel may clog fuel filters by cleaning out your fuel system quickly. If you have an older diesel that has run exclusively on regular diesel fuel for a period of years, it is probably best to stick to B20 blend for your first several fill-ups of biodiesel, and slowly work up the percentage of biodiesel in your tank. It is probably also a good idea to install an additional fuel filter with easily replaced filters in an accessible place, as close to the fuel tank as possible. For owners of older Mercedes Benz models that are likely to have significant amounts of contaminants in the tank from years of regular diesel oil use, check your owner's manual to see whether you have an in-tank fuel filter. If so, look into having it removed before switching to high concentrations of biodiesel.

2.  Biodiesel typically has a higher gel point than regular diesel fuel. That means it pours slower in cold temperatures. As this can be an issue with some petro-diesel fuels as well, there are a number of after-market devices that can help. Failing that, B20 blends are typically indistinguishable from regular diesel in terms of cold weather operation.

3.  In a very few, older diesel vehicles, some of the seals are made of materials that can deteriorate when exposed to high concentrations of biodiesel. Impervious replacement seals are generally available, if required. This is definitely not an issue if the vehicle is rated for use with low sulphur diesel (LSD or ULSD). Also, it is generally accepted that this is not an issue with blends up to at least B20.

## LPG and CNG

Vehicles that run on liquid propane gas (LPG) or compressed natural gas (CNG), sometimes as a dual-fuel capability, are also available on some models as special orders. Fuelling stations for these fuels remain rare, and both fuels appear to be dropping in popularity over time. While the suppliers position these as alternate fuels, they are still fossil fuels. They just are not as mainstream as gasoline and diesel.

Many CNG vehicles are dual-fuel vehicles. In this case, you have the option of switching back to gasoline easily. This flexibility entails a compromise. CNG and LNG could be used in higher compression engines, however, the need to also accommodate gasoline means that lower compression engines have to be retained. Consider the use of ethanol-blended fuels (gasohol) instead of regular gasoline.

If you use CNG vehicles for their environmental benefits, consider switching to another environmentally friendly choice when it is time to replace your vehicle. These could include one of a variety of biofuels (e.g., biodiesel, straight vegetable oil, ethanol, methane).

## Electric Vehicles

Electric vehicles (EVs) provide an even more environmentally friendly option. Regrettably, highway-capable EVs are not available from the major automakers in North America any longer, despite significant advances in their usability in recent years. After flirting with the market for highway-capable electric vehicles as a result of requirements imposed by the California Air Resources Board (CARB), all automakers in North America have scuttled their programs. In most cases, the manufacturers are retrieving the few EVs they did allow into public hands as quickly as possible, and crushing most of them in the hopes that the average car buyer will forget they ever existed. Therefore, you cannot buy a new EV today in North America from a major automaker at any price.

There are still some small operations that will convert an existing vehicle to electric power. Unfortunately, due to the limited numbers of conversions, component prices remain high, and one-off labour costs deter all but the most environmentally conscious from EVs today. Hobbyists and small companies continue to push the envelope with EVs, enhancing the technology (performance, range, reliability), if slowly. Some vehicles do come available second-hand on occasion, but it requires knowledge and patience to acquire one that is a good fit to personal requirements and is affordable. Even some of the CARB vehicles become available second-hand on rare occasions.

Enthusiasts continue to build their own EVs, or convert existing fossil-fuelers to electric power.

One quiet market niche where EVs are thriving is the low speed vehicle (LSV) or neighbourhood electric vehicle (NEV) market. These are EVs that travel at 40 km/h (25 mph) or less. They are suited to urban driving and short missions such as errand running and commuting, the bulk of driving trips for many of us. They are intended for use on residential and low-speed roads. In a previous generation, a similar class of vehicle was familiar in small-town California, known as electric shoppers or run-abouts. In small towns, gated communities, retirement communities, tourist areas, and other suitable markets, these LSVs or NEVs, sometimes denigrated as high-end golf-carts, are getting the job done. This keeps gasoline vehicles from being used for many short trips, where they pollute the most. It also saves the gasoline cars from the trips that are the hardest on them, so they last longer and have a higher resale value.

Still, it is too early to consider the on-road EV truly dead. With the rising number of LSVs being purchased, a die-hard group of enthusiasts still converting and building their own highway-capable EVs, and continuing advances in battery technologies (desired for hybrids, including fuel-cell hybrids), the day of the modern, high performance EV may yet arrive within a decade or so. The waiting lists for the CARB EVs were strong evidence that demand for a competent, affordable, zero emissions vehicle was there.

Another niche market for electric drive is in boats because more lakes are imposing no-motor rules to protect the environment (electric drives are usually acceptable).

Author's electric boat (note solar panel for charging on rear hatch)

## Alternate Modes
## Telecommuting

Telecommuting is more than simply working from home with the benefit of telephones, fax machines, and computers (teleworking). It is any time you can use the telephone or a computer to save yourself a physical trip. It could be comparison-shopping for an item by telephone (remember "let your fingers do the walking"?). More in vogue today is using the Internet to compare prices, participate in auctions, or carry out consumer research. Courtesy of the Internet, savvy car shoppers know more about the vehicle that interests them than the salesperson before they hit the showroom. Virtual tours permit you to check out your hotel room before you leave home. E-mail has largely supplanted regular mail for non-business inter-personal communications due to higher speed, convenience, and reliability. So far, these advantages outweigh the issues of spam, viruses, and extra cost (Internet service) at the expense of traditional postal services. Many Internet users will shell out even more money for high-speed connections, especially as more services become available (e.g., e-mail your digital photos to the store to have them printed as snapshots for pickup later, or order your groceries for home delivery).

As webcams and virtual private networks become more prevalent, it will become easier to work as distributed teams—videoconferences from each member's home or office become practical with very affordable equipment and high-speed connections.

As wireless fidelity (Wi-Fi) extends the reach of high-speed connections beyond the end of optical fibre and wires, high-speed connections will reach wider areas in the future—much as cellular telephones have leapfrogged telephone service penetration into parts of the world that never developed a wired telephone infrastructure.

Teleshopping saves you the time you would otherwise spend in physical travel (stuck in traffic, walking from store to store), but also permits you to shop when the stores are physically closed. Teleshopping also allows you to avoid the annoying part-time sales associates who feel they are making a sacrifice to ask whether they can help you, then quickly prove they don't know enough to be of any real assistance.

## Walking

Take a walk! Seriously. Find ways to displace your travels in cars and even on public transit by walking, when it makes sense and fits into your day. Instead of driving to the nearby convenience store for a couple of items, walk over and back. It might take a few minutes more, but the exercise will do you good, and leaving your car parked will benefit our environment (and the vehicle's resale value).

Walking is cheap. You can get a lot of miles out of a reasonable pair of walking shoes.

Perhaps the most amazing thing about walking is the change of perspective that comes from the change in pace. As a society, we talk about taking time to stop and smell the roses. Well, you can't do it at 100 km/h on the highway. You can do it when you're walking down the sidewalk or along a residential street. You can also chat with a neighbour, hear a bird singing, and otherwise observe bits of nature in your surroundings.

Unfortunately, we can't take walking for granted in our cities anymore. In too many cases highways have destroyed walking routes in favour of high-speed automotive traffic. In our own neighbourhood recently, we had to petition to keep a footpath at the end of our residential street to the commercial area a few hundred metres away. The alternative, by paved streets, would be approximately a 2-kilometre loop to reach the same point. A 4-kilometre round trip would put many residents back into their cars. That's a trivial example compared to the work that has been done since 1945 to design North American residential communities around the car.

Many writers mourn the loss of our villages, consumed by growing cities, urban sprawl, and cookie-cutter big box malls. The fact is we can redevelop our villages within our cities and suburbs, if we choose to do so. In reality, what makes a village is a sense of community, of knowing the people in a small geographical area as a result of repeated interactions. The physical act that makes that happen is walking. Once you get in your car, it doesn't really matter if the service you are looking for is two kilometres away or twenty—it makes very little difference time-wise. Supporting businesses within walking distance will improve your community, your village. If you don't support them, they disappear. That's voting with your dollars and your feet.

## Cycling and Human Power Vehicles (HPVs)

Ride a bicycle. Good, reliable, used bikes are available from bike shops, sporting goods stores, second-hand shops, classified advertisements, freecycle, community organizations, and other sources. You don't need a brand-new bicycle with all the gadgets—you need inexpensive, reliable transportation. If you only do one thing to reduce your transportation energy consumption, in my opinion it should be making sure you have a serviceable bicycle and using it when appropriate.

A bicycle is probably the most efficient means of transportation at our disposal, in terms of distance traveled for energy expended (including energy to manufacture the vehicle). Essentially the only emissions that come from cycling are the carbon dioxide exhaled and heat produced by the cyclist, and those are trivial.

In some circumstances, bicycles or other HPVs may be practical for some trips, especially where bicycle lanes and paths are available. In rush-hour gridlock and traffic jams bicycles can often be faster than cars because they can manoeuvre around cars trapped in traffic and can use a mix of regular roads and pathways that do not permit cars. Hence, the success of bicycle couriers in our cities.

Don't be a hero. If you haven't used a bike for a few years, ease into it. Don't start with a marathon trip. Start with something short, and work up to longer distances and more challenging missions. This is an annual event for me, as I am not a winter rider. Use an electric-assist bicycle if this makes using a bicycle more practical for you.

The author's electric-assist bicycle
(motor in front hub, battery on top of rear pannier rack)

Whichever type of bike or HPV you choose to use, it will improve the air in your locale, take up less space than a car on roads and in parking lots, provide you with exercise and a fitness benefit, and save you money on operating costs.

For your own safety, please invest in a bike helmet and use it. Also, invest in carriers like racks, baskets, panniers, saddle-bags, and back packs to keep your hands free for riding, signaling, and control. There are courses on how to ride safely with motor vehicle traffic and help improve your confidence on your bike if it has been a while since you have ridden seriously.

## Public Transit and Commercial Carriers

Use mass transit when it is practical; e.g., for commuting or attending popular events where parking and traffic jams are often a problem.

Mass transit is a viable option for many of us, and some transit systems are starting to embrace the power of multi-modal travel making them far more effective. Park and ride systems permit people to travel from home in low density areas to a depot where they can connect with the mass transit system, making the transit system more effective. Once into the higher density area, the transit system is often faster than the private car, as it can bypass congested traffic on its own private tracks, roadways, or high-occupancy vehicle (HOV) lanes. Some park and ride facilities actually cater to EVs, NEVs, and HPVs as part of an effective, environmentally superior transportation solution. Some transit systems cater to bicycles. In Ottawa, OC Transpo's Rack & Roll system allows cyclists to load their bikes on the front of the bus for the transit trip, so that they can cycle at both ends of the bus ride. Kingston has similar equipment that can be left on the buses year-round.

Take the train instead of a commercial airline flight for trips up to 400 km (about 250 miles) when this is practical. Train fares are usually quite a bit cheaper, and the difference in time from origin to destination is usually small, even with the slow trains common in North America, for trips up to this distance. Surprised? Air travel is not especially fast for short distances when you consider the time to travel to the airport (usually outside the city, while the train station is usually closer to the city center), checking your luggage at the airline counter up to an hour before flight time (it usually stays with you on the train), passing through security, waiting to board the plane, waiting for your bags on arrival at the destination airport (bags are still with you on the train), and then get from the airport to where you actually want to be (train station is probably closer to your real destination than the airport is).

The train uses less fuel per passenger mile, producing less pollution. Some trains are even electric, producing no pollution while moving you from place to place. There is no cabin pressure problem (that makes your ears pop) or increased radiation exposure on the train.

Given the extent to which North American society (government, corporations, and individuals) have made mass-transit alternatives difficult to access, inconvenient to use, and generally a second-class option, it will take a long time to change general attitudes toward public transit in the U.S. and Canada. However, the more we use public transit, the better it will get.

## 25.1. Indirect Transportation Usage

There is a saying in the trucking industry, "If you bought it, a truck brought it." Obviously, transportation of goods extends beyond trucks, and includes ships (barges), rail, airfreight, and other modes. Most of us have no idea how much energy is spent on the transportation of the goods we purchase. It is enormous, and growing, especially as industry and retailers have embraced the concept of reduced on-site inventory (especially downstream in the delivery chain) and just-in-time (JIT) delivery. Again, I am not advocating that we all stop buying the products that we need or enjoy. I am suggesting that you consider the transportation energy implications (energy use, environmental impacts) of buying something produced on the other side of the planet, especially if a suitable alternative is available that is produced closer to home. This may be one of the few times it is appropriate to compare apples and oranges, to put a new spin on an old phrase. In this example, both are fruits, and both provide nutritional value (notably fibre and vitamin C). However, they are typically not grown in the same locales, as the trees that bear them prefer different climates. By extension, the same logic applies to their derived products (e.g., juices).

## 25.2. Local Sourcing

One way to reduce our indirect energy consumption is to purchase things produced closer to our location, so as to reduce the associated energy required to transport it. In general, the further it travels, the more energy is used to move it. This will be true until such time as carriers start to make use of less energy-intensive shipping technologies; e.g., sail-assisted ships, hybrid trucks, electrified rail.

One topic that has brought a lot of focus to this specific issue is food-miles. In short, how far has the food that we eat traveled in order to reach our mouths? In many cases, the energy consumed to transport food to us far exceeds the energy value in the food we actually consume, even ignoring issues of spoilage and other waste. This is especially true of exotic perishables, which can include fruit that is out of season locally, which may make use of more energy-intensive (i.e., faster) modes of transport in order to reduce transit time (and spoilage). While the food-miles label has the appeal of simplicity, it might be more appropriate to think of associated transport energy, as shipping of preserved fruit by ship, train, and truck may be considerably less energy-intensive than air-freight per food calorie delivered to the plate. This approach has the disadvantage that it will be considerably harder to calculate with any degree of accuracy.

Still, it is a reasonable assumption that crops grown closer to us will require less energy for transport than products that arise further afield. To return to our example of comparing apples and oranges, if we live in the northern U.S. or Canada, apples may be a local crop, but oranges are almost certain to travel at least 1,000 kilometres (over 600 miles) to reach us.

The long-standing and wide-spread practice of subsidizing energy prices in North America means they cannot be relied on to tell us the real cost of the energy imbedded in products and their implicit transportation. This has made it easier for the global corporate web to export jobs from North America by subsidizing the combined price of offshore production plus transportation.

There are several means of reducing the amount of energy used in transporting our food. Here are a few.

## Household Gardens

There are lots of reasons to have your own garden, however small. It provides fresher produce, with higher nutrient value, than you can buy at any market. It gives you absolute certainty with regard to what nutrients (fertilizers), herbicides, and pesticides (preferably none) have been used on this food you will be consuming. It gives you a connection to nature and its cycles, and challenges faced by those that nurture the other food you eat. It provides you with some exercise, and an excuse to spend some time outdoors. However, in my opinion, none of these compare with the sensual experience of eating truly fresh produce that was still on the plant just hours, or even minutes, before. Words on the page simply cannot express the difference between peas or beans fresh from the garden and those coming from cans or the freezer. They simply are not the same. Until you have had a tomato sandwich for breakfast that includes a tomato that was on the vine less than half an hour earlier, you cannot conceive what a difference that makes.

Many who have not had the benefit of this transforming experience are under the impression that they need a large plot of land to have a vegetable garden. That is simply not the case. It takes very little space to grow something. It's best to start small anyway—don't turn gardening into a chore—try to keep it fun and interesting. If you have some lawn with good sun, turn over a small section (say one metre by two—a couple of square yards) in the fall. Add compost (if you don't make your own, you can buy some) in the spring, and buy seeds for a vegetable you like (yellow bush beans, squash, beets, radish, cherry tomatoes). Plant according to the instructions, and tend to the plants once or twice a week (mostly weeding and watering). If you don't have that much space, look into window box or container gardens (which have the advantage of mobility, but likely need more watering), or even guerilla gardening if appropriate space is nearby, but not actually yours. Guerilla gardening is the practice of using vacant, unmaintained land to grow food crops. You can expand your garden in subsequent seasons if you so desire.

Use compost generously instead of commercial fertilizers—you'll be amazed by the difference in your plants. Mulch and weed instead of using herbicides. Healthy soil and crop rotation will eliminate the need for gardeners to use pesticides.

Warning: gardening can be addictive. Once you taste a tomato that was on the vine just minutes earlier, or leaf lettuce just picked, washed and made into your salad, you may look upon the store-bought variety rather differently. The same applies for most fruit and vegetables. Once you have felt the satisfaction of combining soil, seed, water and sunlight to produce food, and partaken of it, you may be hooked (not for life, but on life). You may learn about new and different varieties of tomatoes, green beans, yellow beans, beets, apples, and you can have something other than the single variety of each offered by your local supermarket. You may learn there are heirloom or heritage varieties of vegetables better suited to your locale than the seeds offered by your local hardware store or even a distant seed house. You have been warned.

While the shadow of genetically modified organisms (GMO) has not become an issue for the home gardener in North America yet, it is probably not far off. The government of Canada has taken a position that could render illegal the personal practice of collecting and using your own seed. Which is only slightly more absurd than the continued practice of using GMO seed in the industrial agriculture field. I'm no expert in the field, but the independent research I have found on the subject indicates that GMO seed does not produce the anticipated benefits of reduced need for weed suppression or fertilizing the crops. If the touted benefits are not being realized, why do farmers continue to pay the higher prices for GMO seed, and accept the risks attending the use of GMO crops? They could possibly face stronger weeds and lawsuits by GMO producers if GMO strains are accidentally seeded onto someone else's land.

I understand that gardening is not for everyone. We don't all have green thumbs, and I still don't know how to take the tedium out of weeding, although some serious tools (like our electric tractor with tiller) definitely help. Still, most of us can grow something edible (even if it is just leaf lettuce or radish).

Electric tractor and tiller in the author's garden

For those vegetables that simply defeat us (in my yard, it's cucumbers), the local farmer's market is a good option.

## Buy Locally

There's a reason humans established currency. Some of us have a natural advantage in doing certain things (e.g., growing cucumbers) relative to others (e.g., me). So, let's benefit from these differing talents and exchange our personal bounties for our mutual benefit. Most areas have small markets where local producers exchange their harvest for mere money. Even if it doesn't come from my yard, a local cucumber is better than no cucumber at all.

## Buy In Season

However, just because a fruit or vegetable is available at the supermarket or the farmers' market, doesn't mean it really belongs there in terms of nature's cycles.

If you just have to have fresh strawberries in February, despite the snow on the ground outside, well, the market is there to serve you. Just be aware the amount of energy spent to deliver that berry to your table dwarfs the food energy value it represents.

One variant (recent in my experience) that has the potential to overcome the issue (out of season local produce) is local commercial greenhouses. There is an operation about 25 kilometres from my house that produces tomatoes year-round, despite snow on the ground and temperatures that can drop below -30°C. This is accomplished by an extensive investment in greenhouses that operate all year. I have no idea what their energy bill is, but given my experience, with good design and appropriate siting, year-round greenhouses should be feasible with only minimal supplemental energy beyond what solar heating and thermal mass can accomplish.

## Preserving

There are many options for preserving food beyond refrigeration and freezing. These include: drying, pickling, canning, salting, smoking, spicing, cold storage techniques, jams/marmalades, and more. You can also substitute more durable items for more perishable items (e.g., some varieties of apples instead of citrus fruit).

There are many good books and Web sites that cover these topics in detail, so there is no need to take up the space here.

## Food Co-Ops (Farmers, Gardens)

If you don't have the time, space, or inclination for a garden, but still want its benefits, look into the potential for a co-operative arrangement between a local farm and/or other families. Such arrangements can include providing some labour during planting and harvest times (and perhaps some occasional weeding), to simply financing a portion of the expected harvest or renting your own plot of land. Much of this kind of activity is starting to be called Community Supported Agriculture or Community Shared Agriculture (CSA). As this is an area that is in much flux just now, and continuing to grow, it is difficult to provide good coverage in a static document. For now, in the U.S., I recommend visiting the Alternative Farming Systems Information Centre CSA Web page as a starting point:

http://www.nal.usda.gov/afsic/csa/

For Canadian CSA farms, try the Biodynamic Farming and Gardening Association's page at:

http://www.biodynamics.com/canada.html

## 25.3. Packaging

It takes energy and materials to make packaging. And much of the packaging we buy (yes, we pay for it) is discarded and becomes a waste management issue. It adds to the bulk we carry, the weight we move, the cost of the products we buy, and the cost we pay to dispose of waste. This includes even the shopping bags we carry from the store. The grocery store closest to us actually charges five cents per bag they have to supply. They say that reflects their cost for the plastic bags that are derived from oil products. That's a real motivator for many to take their own bags. Given we already do, it's a motivator for us to shop there. We save anywhere from a dime to a dollar on each trip because of our durable shopping bags.

There are other opportunities to re-use your packaging and containers. For example, the local gas station (that carries ethanol blend fuel) also dispenses bulk windshield washer fluid in your existing container, which many of us carry around in our vehicles anyway.

Many household products now come in refill versions, as well as the original container, that may double as a dispenser. In some cases, you can provide your own containers for bulk purchases.

Even when you can't avoid the packaging in the initial purchase, sometimes the packaging can be re-used for other purposes. Detergent spray bottles can be cleaned out and used to spray water onto household plants, or vinegar and water solutions for cleaning glass, etc. The possibilities are limited only by your imagination.

## 25.4. Embedded Energy

Many of the products we use require a large amount of energy to produce. For example, metals typically have to be mined, refined, smelted, stored in standard form (ingots, bars, sheets, etc.), and then re-formed (moulded, cast, machined) to their desired form. All of these processes use energy and have environmental impacts. Glass, plastics, electronics, and many other products we see in our everyday existence are also the result of energy-intensive processes. In general, everything that is closer to its original state represents less embedded energy used in its production. So, it is worth considering exactly what it is we are buying in terms of the raw materials when we make our purchases.

## 25.5. Waste Management

How far does your garbage and other waste travel? I have not been able to find figures for waste miles as are now available for food miles, but I have come across a few statistics of interest.

In 1998, the U.S. Congressional Research Service determined that 28.4 million tons crossed state lines to reach a disposal site. That was over 13% of the residential waste in the U.S. that year.

The City of Toronto trucks its garbage to Michigan, in the U.S. for disposal—about 20,000 truckloads annually. In the meantime, cities like Ottawa are actually reducing the items covered by their curbside recycling programs, so the amount of household waste going to landfill is actually increasing. It costs somewhere between $500,000 and $1,000,000 per acre to establish a landfill, just to fill it with our garbage. Before you make a mountain out of packing peanuts (Blackstrap Mountain in Saskatchewan was made from garbage), just think before you throw. Can the item be re-used? Could it be repaired? Could somebody else make use of it? Can it be recycled, even outside a municipal waste program? For example, many private shipping/mailbox operations will happily take your packing materials—it saves them buying more.

You can find more waste diversion ideas at:

http://www.econogics.com/en/enreusea.htm

Freecycling is another relatively new means of distributing goods you no longer want to people who can use them. Of course, it also works in reverse—you may be able to obtain something you need or want for free. For more information, or to find a local freecycle group, check out the Freecycle Web site at:

http://www.freecycle.org/

Author at Freecycle promotion event

The less you buy (especially packaging), the less you will have to throw away. The less you throw away, the less energy will be spent on picking it up, compacting it, hauling it, dumping it, covering it, and monitoring it for the coming decades. Finally, the less you throw away, the less new landfill space we need to make, and the less your tax bill will have to cover.

Garbage may be worthless, but it's still expensive.

# 26

## *Stationary Energy Use*

This section is organized primarily by energy use, so that you can compare various options for a specific change more easily. It will also make it easier to reference when you have finished reading the book. Retrofits are often caused by necessity; e.g., the fridge broke down—what are our options? By presenting the information this way, you can find other options that might be viable substitutes. Also, by reading through this section in advance, you can think about changes to infrastructure in advance to plan for future substitutions. For example, how can I reduce the energy used by my current natural gas clothes dryer? Options might include a clothesline, or a tumble dryer that uses solar heat as a pre-heater or a substitute for the natural gas heating. However, these options may not be easily implemented in a matter of hours. It may take time to plan and install a pole to support a clothesline, or solar heating panels to feed a clothes dryer.

A great deal can be accomplished in the area of energy savings in the residential sector when starting with the design and construction of a new building; the proverbial clean slate. The opportunities available (siting, using solar energy, geothermal energy, thermal mass, daylighting, super-insulation, etc.) could easily fill another book. In fact, there are a number of books on the subject already.

However, the reality is that few of us will have the opportunity to design and build our own homes, let alone with energy conservation as a primary objective.

Canada has had a superb standard for energy efficient homes since 1981, R-2000, yet only a small fraction of homes built since then have attempted to meet those goals.

As homebuilders (and only to a slightly lesser extent, homebuyers) continue to focus on initial build/purchase price over maintenance and operating costs, even new homes are unlikely to incorporate much in the way of energy saving features beyond the minimums in local building codes.

Housing stock in North America has an average life of over fifty years. So the turnover is something less than 2% annually. Given that the majority of the housing stock that will be in use by 2030 (25 years or another human generation into the future) is either already built or will be built to current (or near-current) standards, I'm going to focus on changes that can be accomplished by behaviour or retrofit and that do not depend on architectural changes. Besides, we simply don't have the physical resources or the cheap energy remaining to undertake a moon-shot program to rebuild the majority of our housing stock (let alone commercial buildings) in the next decade. Realistically, if we want to make a difference, we have to look to upgrades and retrofits as a major piece of the eventual solution to our profligate use of energy today in North America. Even if we built every new home to super-efficient standards starting today, a decade from now, over 80% will still be out in the cold, figuratively if not literally.

Our stretch goal for our household energy conservation activities should be what is sometimes referred to as a Zero Energy House (ZEH). This means a structure that consumes zero net energy in operation. This is not often achieved as an economical retrofit; it isn't even often achieved in new construction. One class of ZEH buildings is off-grid homes. The requirement for self-sufficiency imposed by the lack of a grid connection often leads to efficiency and innovation.

It may not be practical for us to actually achieve a ZEH. However, the concepts embodied in ZEH structures, like off-grid homes, can provide us with valuable guidance as to what works and what does not.

I have read from several sources that the primary energy consumers in the typical North American home are climate control (heating and cooling, approximately 60% of energy used), hot water (20%), appliances (15%), and lighting (5%). My own experience suggests the order of importance is correct, even if the actual percentages vary. Remember, this is total energy, not just electricity, so lighting may well represent more than 5% of your electrical bill. I will approach these four major areas in this order. That doesn't mean lighting should be your last priority, as it presents an easy win for most of us, with a great financial payback.

I want to emphasize we don't have to accomplish this overnight. We are at Peak Oil, energy costs are going to continue to rise, but we still have a few years of relatively cheap energy ahead of us to help with the transition to a much more efficient infrastructure for our future.

For those of you concerned about greenhouse gas (GHG) emissions, I have tried to establish energy costs and savings on a consistent kWh basis. How that corresponds to actual GHGs produced depends entirely on the energy source you are using. If your energy source for electricity is wind power or hydro, the GHG emissions are essentially zero. If your energy source for electricity is a forty-year-old conventional coal-fired generating plant, that's an entirely different story. Similarly for heating fuels, the GHGs produced vary across the spectrum from solar, to biofuels, to electricity, to natural gas, to heating oil. If you don't want to do your own research, here are some numbers you can work with.

For heating, natural gas is the current leading choice for new construction and retrofits despite dwindling North American reserves, so I'll provide figures for that here. The standard measure for consumption of natural gas in most of North America is a thousand cubic feet (mcf), the equivalent of a box measuring 10 feet on each side. It's the equivalent of 10 therms, or 1,028,000 BTUs, or about 28 cubic metres, or approximately 300 kWh. For those of you using electricity, oil, or wood, hopefully those conversions will give you an equivalent sense of scale.

Whatever way you measure it, burning that mcf of natural gas will result in the emission of about 180 kg (400 pounds) of carbon dioxide. So check your gas bills to establish your consumption, then multiply accordingly to figure out your GHG emissions. This figure does not include emissions resulting from the exploration, drilling, refining, production, storage, or transportation of the natural gas, only the burning of it at point of use.

For electricity to run our appliances, I'm going with the figure of 0.23 kg of $CO_2$ per kWh of electricity consumed (per a document on the David Suzuki Foundation Web site). The U.S. EPA uses a figure of 1.64 pounds of $CO_2$ per kWh, approximately .75 kg, which likely reflects the much higher use of coal for electricity generation in the U.S. than in Canada. Most of us are familiar with the kWh as the unit of consumption for electricity.

Coming up with an unassailable figure is almost impossible given the North American mix of coal (just over 50%), nuclear fission (approximately 20%), natural gas (about 16%), hydro and all other renewables (almost 11%), and oil (about 3%) used to produce electricity, and varying degrees of efficiencies at the various individual plants.

For more ideas how to reduce your energy consumption, I also recommend the Energy Saving Now Web site at http://www.energysavingnow.com. It provides tips and tricks in a number of areas, including building envelope, water systems, lighting, heating and cooling, and more.

## 26.1. Household Climate Control

Most people are in the habit of thinking in terms of household heating and air conditioning rather than household climate control overall. However, that is focusing on a specific solution, rather than thinking about what is really desired, and opening our minds to different solutions. In reality, for most of us, what we really desire is a comfortable temperature, probably in the range of 19° to 26°C (68° to 80°F) in our homes, so that we can move around without needing bulky clothing to stay warm or sweltering in unbearable heat. This is a complex subject, as it involves more than just temperature, but also humidity, level of physical activity, cultural norms regarding dress, air flow, heat striation, characteristics of specific technologies and implementations, etc.

Rather than try to treat all of that in this volume (there are other books that cover these subjects), we will assume that most of this is fixed in your current home, and the changes we can make fall short of demolishing your current house and replacing it with a live-in energy experiment showcase.

However, as the root cause of over half of our energy consumption in our homes, and usually over half of our energy bills, climate control is worth some consideration. I am not proposing that we freeze or sweat in the dark. I am proposing that we take some common sense steps to reduce our energy consumption, our dependence on diminishing fossil fuel resources, and our personal energy bills.

There are some steps we can take that will reduce both our heating and cooling demands, and some that are specific to one or the other. I will cover those items that help with both aspects first (double gain), and then those that are specific to space heating and cooling (air conditioning) under separate headings after that.

## Weather-Sealing and Insulation

In my opinion, the best solution to excessive energy consumption is conservation. No matter what fuel you end up using, conservation should enable you to use less of it. By reducing demand on whatever heat source you do use, it should cost less to operate (fuel) and last longer (reduced maintenance and replacement cost).

Even after years of being told this is a prime target for energy savings, improved insulation and weather-sealing is still the number one opportunity for major energy savings at home for most of us, for heating and cooling. Look for drafts, and seal them off. Use expanding foam, foam strips, wall outlet and switch gaskets, plastic sheeting, door sweeps, knob covers, and weighted fabric rolls at the bottom of doors, etc. Use the thin foam strips to help seal gaps around doors. Your local hardware store can supply all of these, and probably more ideas.

Is there enough insulation in your attic? Use up to R-40 in ceilings in most of Canada and the northern U.S. (conventional standard), if there is room. That's about 30 cm (12 inches) of fiberglass insulation. This requires an investment that typically pays back in 3-5 years, assuming no increase in fuel costs. It will pay back faster if fuel costs increase. Super-insulate your building to reduce heating and cooling losses even further. R-60 is considered super-insulated by most. R-60 will reduce energy requirements further than R-40, but the payback period will be longer. The R rating system indicates how much resistance the material has to heat loss. Higher numbers are better, indicating higher heat loss resistance. Make sure you do not block soffit openings in the eaves or roof vents.

One advantage electric heat has is that it does not produce carbon monoxide from combustion, so the tighter you can make your house, the better.

Improve weather-sealing to reduce air changes. If your house manages to become too airtight, investigate the use of an air-to-air heat exchanger.

## Thermostats and Timers

If you have a central, forced-air heating system, get a programmable, setback thermostat. During the heating season, set it to lower temperatures when everyone is asleep or out of the house. If you have a regular schedule, and can set the temperature down both at night and during much of the day, you could see savings of up to 30% on your heating bill compared to no setbacks at all. Program the thermostat to return to desired heat level about 30 minutes before people get up or return to the house. If it takes longer than 30 minutes to bring your house back to temperature from the setback level on anything but a record-setting day, you really need to focus on upgrading insulation and weather sealing.

During the heating season, set the temperature as low as you find comfortable. The lower the temperature, the less energy it takes to maintain that temperature. This is because heat transfers from hot to cold proportionally to the difference in the temperatures (sometimes referred to as $\Delta T$ or delta T).

The programmable thermostat is a small investment that will often pay back within the first heating season if you have regular setback opportunities.

During the cooling season, set the thermostat to the highest setting you find comfortable, and dress for the season.

Use the programmable thermostat to cool the house just before you expect to return home at the end of the day, and possibly again before bedtime, if that helps you get to sleep. For window air conditioner units, use a timer rated for the power rating of the unit. It is not necessary to keep your house as cool when no one is home, and it will use less power if controlled by a timer.

If you live in an area that provides interval pricing for electricity consumption (time-of-use metering), you may also wish to use the programmable thermostat or timer to minimize the operation of your air conditioner at peak price periods and maximize its operation when electricity is available at the lower price.

If you have a regular schedule, let the building warm up when it is unoccupied, and set the timer or programming to resume the desired temperature about 30 minutes before you will arrive. If you are going to be away for a few days, shut air conditioners off.

## Thermal Mass

Thermal mass is essentially a means of storing heat (or "coolth") from when it is available until it is desired. It stabilizes the temperature of the building interior over the course of the day or longer. The classic example of this is adobe structures in the U.S. southwest, with their thick walls. The walls absorb heat from the sun during the day, protecting the occupants from searing daytime temperatures. Later, the adobe walls release the heat into the building at night, protecting the occupants from the cold desert night. Less well-known, but still effective are masonry stoves (also known as Russian heaters). These are stoves, usually wood-fired, that are encased in a significant mass of stone, brick, ceramic, masonry, or some combination of those. When the stove is fired up, it heats not only the air in the building interior, but also the mass surrounding the stove. After the fire is banked or goes out, the masonry continues to give off heat for hours, continuing to warm the building. We can apply the same thermal mass concept in our homes. Think of it as thermal inertia, resisting changes in temperature.

In the case of heating, we can use thermal mass to absorb heat when it is available from whatever source. As the temperature falls subsequently, heat will be withdrawn from the thermal mass. Thermal mass can take many forms in our homes, for example, heavy furniture, bricks that support shelves, books, bottles of water secreted out of sight in closets, soil for potted plants, stoneware, ceramic tiles, glass bricks, and other such decorative accent items.

When it comes to cooling, the same thermal mass can be used to absorb "coolth" when temperatures are cooler, and we ventilate the house to cool it off. In areas where temperatures do not cool enough to make that feasible, and interval pricing for electricity is available, the thermal mass can be used to store "coolth" provided by air conditioners when electricity prices are lower, avoiding the need to run the units when electricity prices are higher. The thermal mass provides a means of stabilizing the indoor temperature throughout the course of the day.

Thermal mass in the building interior works best when the building is well insulated and weather-sealed, so there is an effective barrier between the desired temperature inside and the undesired temperature outside.

Thermal mass may reduce the effectiveness of programmable thermostats on your heating and cooling bill to some extent, unless you take the mass into account (slows the heating and cooling of the building).

## Windows

No matter how good your windows are, they are probably your major point of heat loss in winter and heat gain in summer, unless you take additional measures. Very good windows typically have an insulation rating in the order of R4 to R6, which compares poorly even with a minimally insulated exterior wall with 9 cm (3.5 inches) of insulation, house-wrap, wall-board, vapour barrier, and exterior siding at roughly R12.

If you find condensation on windows, then focus on better insulation for the windows (e.g., the plastic covering products), improving airflow over the windows, or replacing the windows with units with better insulating properties.

I see two basic ways to improve the energy efficiency of windows: help them or replace them.

If your windows are noticeably colder than the adjacent walls, or routinely collect condensation or even ice on cold days, they are likely culprits making your heating bill grow.

The cheaper, faster solution is to help your existing windows. This is done by adding layers of insulation and sealing to what already exists.

In some cases, storm windows can be added on the outside to provide an additional barrier to heat loss in winter and reduce heat gain in summer. Storm windows can be installed and removed on a seasonal basis, or left in place year-round.

Clear plastic can be added both inside and outside. There are plastic kits that will shrink with heat (e.g., a hair dryer) to provide a tight seal around window frames on the inside of windows that don't need to be opened in the winter. On the outside, clear vapour barrier or purpose made sheets can be installed and removed seasonally. Fasten the plastic so that it can be easily removed and re-used in future heating seasons. The plastic will usually reduce drafts as well as add a thermal barrier.

Insulated window coverings can also improve the effective R-factor of your windows. This can range from blinds filled with collapsible honeycomb fabric, to insulated drapes, or quilts fitted to the window opening.

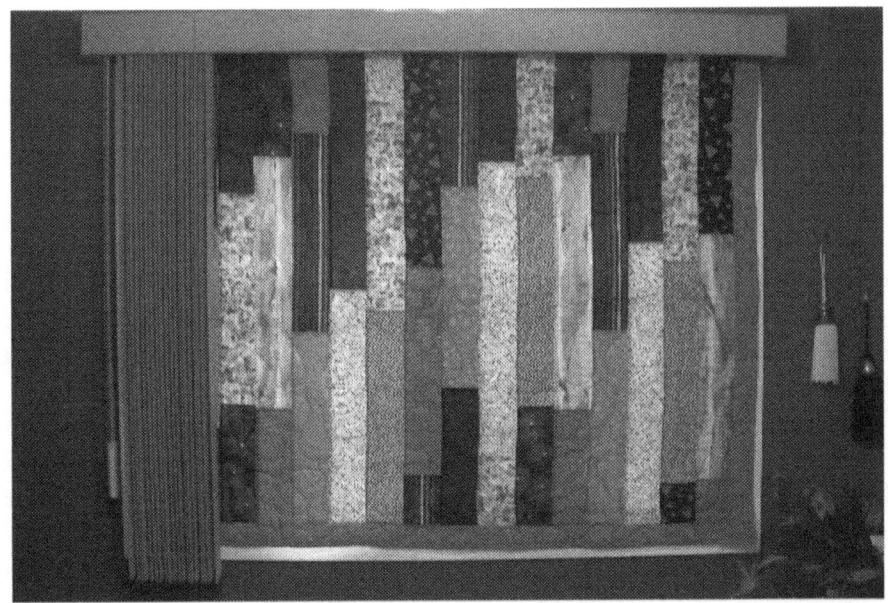

Quilt on window in author's home

If you can afford them, newer, more energy-efficient windows are a better long-term solution. Not only do they dispense with the seasonal chores of putting plastic up and taking it down, they will help you save more energy (and on heating bills) as well. If you do decide to change windows, do your homework first. While low-emissivity (low-E) windows are good, there are some variables to consider. You may want to consider the thermal gain properties of sun-facing windows to achieve the maximum passive solar energy benefit, or reducing the size of windows that never see sun. The original building plans for most homes do not consider orientation of the house, so may well have placed the largest windows on the sunless side of the house, and smaller ones facing the sun. In very cold climates, triple-glazed windows may be worth the additional cost for the subsequent savings benefits. Vinyl cladding is usually better than metal or wood frame windows.

Of course, even energy-efficient windows can benefit from use of additional insulation like the sealed plastic, insulated blinds, or quilted coverings.

## Heat Pumps

Heat pumps are essentially reversible air conditioners. In summer, they work like a conventional central air conditioning system. They cool the inside of the building by extracting heat, and dumping it outside. In winter, they extract heat energy from the outside (leaving the outside air cooler), and transfer the heat inside. Air-to-air (conventional) heat pumps typically work best when the outdoor temperature is above the freezing point of water, 0°C (32°F).

## Geothermal (Ground Source) Connection

If you are installing a heat pump or central air conditioning in new construction anyway, consider the use of a geothermal or ground-loop coupled system. The ground is typically much cooler (about 12°-15°C or 55°-60°F) than outside air when air conditioning is required. Using this source of "coolth" means your air conditioner will not have to work as hard cooling your building on hot days.

In winter, when outdoor temperatures fall below the freezing point, reducing the effectiveness of conventional heat pumps, you can extend the period for which your heat pump will be effective by drawing warmth from the ground below the frost line.

If the ground area you are using as a heat-dump in the summer has some degree of insulation (depending on the type of covering), it may also provide some additional heat energy for your heat pump during the heating season. Conversely, if you have cooled this area during the winter by extracting heat energy, this zone may provide additional "coolth" during the cooling season. A version of this is being implemented in Okotoks, Alberta, where large amounts of solar heating energy are being pumped into the ground in summer as a heat reservoir, to be extracted again in the winter to help heat an entire residential sub-division.

## Reflective Control Coatings

Reflective Control Coating (RCC) is a relatively new entry in the arsenal of energy-saving technologies. These coatings are typically elastomeric compounds that can be applied much like paint, but which reflect heat energy as a radiant barrier. There are types that can be applied to both interior and exterior surfaces, including roofing materials. In general, it appears that if your primary objective is to retain heat, it is most effective to apply the RCC to interior surfaces. If your major objective is to reduce heat gain (e.g., solar gain), then it is more effective to apply the coating to the exterior, either roof or walls, or both.

To date, people trained in their use usually do the application of these materials, not end-customers. There do appear to be some issues in the market with some inferior materials being sold by less reputable suppliers. Do your research before deciding on a brand and supplier.

Results from using RCC appear to be somewhat variable, probably due to the quality of the various products, who applied them, and the specifics of a given installation. There appears to be a reasonable degree of agreement that results will be less dramatic in homes that are already well insulated and weather-sealed.

## 26.2. Space Heating

Space heating is usually the largest single use of energy for most households in colder climates over the course of the year. Increasingly, more households use natural gas for their primary space heating energy source. Other major primary energy sources for space heating in North America include heating oil and electricity. Some homes use wood and other biomass, but this is usually not a primary heat source. There are a variety of other techniques used such as heat pumps (ground-coupled or air-based), active solar, earth houses, and even some still using coal.

Even if you don't heat with electricity, your furnace fan is probably electric. A typical furnace blower fan uses about 700 watts when running. Some use more. The less time your furnace fan is running, the less electricity it consumes. So insulating and weather sealing your house and lowering the thermostat will save electricity as well as whatever heating fuel you are using.

Switching fuels will likely require the acquisition of a new furnace or stove—typically a significant investment. Also, with the volatility of prices for common heating fuels, there is no guarantee that the cheaper fuel today will remain so in the long-term. Therefore, switching fuels does not have the same value as many of the other measures suggested in this section. In particular, switching between different fossil fuels is likely just delaying the inevitable. However, if you need a new furnace in the near future, you may want to consider some of the less mainstream options. Locally available fuels may be a factor (e.g., corn, wood pellets, scrap wood, methane, etc.). If you have the space and funds, I seriously recommend investigating ground-coupled heat pumps as an option. They provide both heating and cooling, and typically a dramatically lower energy bill for both. However, they are frequently not appropriate as a retrofit (cost and disruption).

### Slippers, Socks, and Sweaters

The warmer you are, the cooler your house can be. The cooler your house, the less heat energy it takes to maintain the temperature. Warmer air rises, and in most buildings the air near the floor is colder than the air near the ceiling. So, warmer socks and slippers will accomplish a lot to make us feel warmer. Our rule of thumb—if your fingers are cold, the room is too cold. Otherwise, put on warmer socks, slippers, or a sweater. We also use lap quilts for when we are awake, but inactive for extended periods of time. Use an extra blanket at night, if this allows you to lower the temperature in the room or the rest of the house.

## Use Free Heat

When the sun is shining, make use of it. Open the window covering, and let the sun stream in. Make sure your windows are clean in the fall to make the most of this heat source.

When cooking, use the range-hood exhaust as little as possible so that the heat being produced is not being sucked out of your house. If the kitchen gets too warm, use a fan to move the heat into other parts of the house, including the furnace fan if necessary.

When you have boiled water to cook pasta or vegetables, etc., drain the water into another container (e.g., a large, emptied coffee tin or juice can), and let the hot water sit there until it cools to room temperature, extracting all the heat you can from it.

When taking a shower in a bathtub, put the stopper in the drain. Let the hot water stand in the tub until it cools to room temperature, to extract the heat from it. Then drain the water. Similarly, when taking a bath, let the water stand until cool. Note: these heat-saver steps will likely lead to soap-rings around the bathtub, which may require some extra cleaning effort later.

Use bathroom exhaust vents as little as possible. Their primary function is to remove humid air when you are showering. In winter, you may be able to use that humidity in your house, as well as the heat from the warmed, moist air. Try leaving the bathroom door open a bit when showering to let the humidity escape to the rest of the house. If your exhaust fan is wired to the bathroom light switch, it is throwing heat out of your home whenever the light is on, not just when there is excess humidity. Look into installing another switch to control the exhaust fan separately and keep that heat in.

If you use an electric clothes dryer, you can vent the heat from it into your house during the heating season to get the most benefit from the power you are paying for. Typically, we hang our clothes outside to dry in the warm season, and use an electric dryer when there is snow on the ground. In the winter we use both the heat and the humidity from the clothes dryer in our house; winter cold dries the air here. **N.B. If you use a natural gas clothes dryer, do not vent it indoors**

**under any circumstances—there is a significant carbon monoxide hazard.**
Carbon monoxide is a colourless, odorless gas, and it is deadly.

If you vent the dryer indoors during the heating season to retain the heat and humidity, don't forget to block off the duct hole that goes through the house wall to reduce cold air leakage into the house. There should also be a vent flap on the outside to reduce cold air infiltration, and entry by pests.

A vent switch box makes the seasonal change-over easier, and provides a means of securing the lint sock to capture lint that gets past the dryer's internal filter.

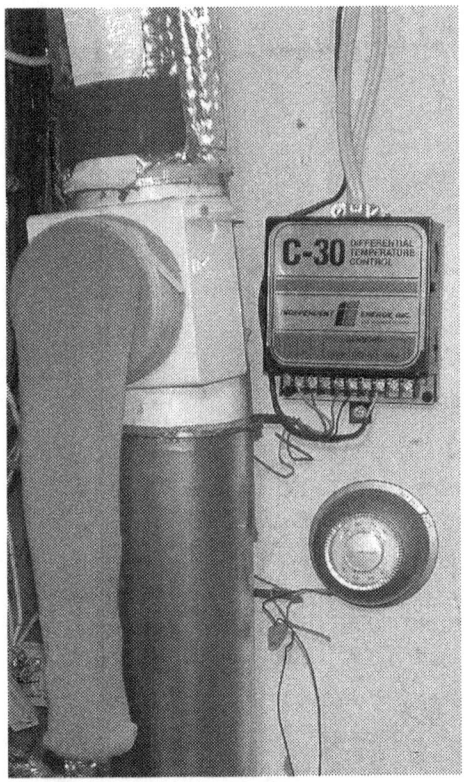

Vent box for dryer with lint sock (left)—also shown is differential
thermostat for solar heating system.

If you don't want the humidity, but do want the heat, look for a condensing clothes dryer. These are still rare in North America, but apparently they can be obtained with perseverance.

## Closed Door Policy

If you have rooms you do not use, reduce the heat in them and close the doors. If you have a central, forced-air system, close the registers to these rooms. If you can choose which rooms to close, choose rooms that don't face the sun rather than sun-facing rooms. If you have baseboard heaters, set the temperature in these rooms to about 5°C (45°F). To avoid mould, make sure there is some air circulation and hopefully direct sunlight on a regular basis.

Keep exterior doors closed as much as possible. Use air-lock systems where possible, such as vestibules with inner and outer doors.

# 26.3. Space Cooling (Air Conditioning)

Air conditioning is probably the biggest single load faced by the electrical grid on hot, summer afternoons. Air conditioners create the largest peak demand for electricity in North America, resulting in the highest prices for electricity. The more you can reduce your demand for electricity on hot summer days, especially in the late afternoon, the lower our overall bill for electricity will be.

There are things you can do. As noted above, the first thing is to maximize your thermal barriers (insulation and weather-sealing). The harder it is for heat to leak in, the less work the air conditioner has to do.

## Reduce Solar Gain

During the cooling season, the biggest source of heat in most residential buildings is solar gain. The best way to combat this is to stop the sunlight from reaching your windows. Sunlight that does not reach your windows cannot be converted to infrared heat energy that heats up your house. Use awnings, shutters, roof overhangs, trees, and other coverings to shade your windows. There are also some reflective films that can be applied directly on the glass to reduce the amount of light that enters through the windows. Shade cloth is another covering meant to be applied on the outside of the windows to reduce solar gain. This is far more effective than using indoor window coverings that allow the heat to enter your home, but try to trap it against the window. If you cannot block sunlight outside the sun-facing windows, try to use reflective window treatments on the inside so that at least some of the heat energy is sent back out through the glass rather than being absorbed in your home.

Shade cloth on windows on author's home

Suppose a house has 8 square metres (9.7 square yards) of sun-facing windows (likely a low figure for the average Canadian house), with no shade or outside covering. Allowing for the angle of incidence and some ground reflection, let us assume that this is the equivalent of 6 square metres (7.3 square yards) for direct solar energy (insolation). Let us further assume that the double-pane glass typical of most Canadian buildings converts only 50% of that energy into heat as it passes through the glass. That's still 3 kW of heat! (8 square metres x 1 kW/sq. meter insolation energy x 75% incidence/reflection allowance x 50% conversion factor.) Suppose you get the equivalent of 7 full-sun hours on a sunny summer day. That's 21 kWh of heat coming in through your windows during the day. Put another way, it's like taking two of those 1500-watt portable electric heaters, and turning them on at full power in your house for 7 hours on a day that is already hot.

So now, you need your air conditioner to counteract that heating effect. Assuming a Coefficient of Performance (CoP) rating of 2.0 (typical of a window unit or low-end or older central unit), then 10.5 kWh of electricity will be needed to get this performance from your air conditioner. At approximately

$0.15 per kWh, that's just over a dollar-and-a-half a day. However, consider that this electricity will be required at the absolute peak demand time, when everyone else's air conditioners are also running. During the summer of 2002, electricity was selling in Ontario for as much as Cdn $4.71 per kWh on the spot market at peak demand time. If you had to buy your 10.5 kWh at that price, the cost to you would be $49.46 per hot summer day. Would you be prepared to think about shading your windows to avoid that sort of expense showing up directly on your electrical bill? At that price, the cost of awnings for the entire sun-facing side of your home would probably be recovered in a single summer. However, even at a price of approximately $0.15 per kWh, some sort of summer shading for sun-facing windows is still probably a worthwhile investment. Assuming 10.5 kWh per sunny day to be saved, and up to 60 such days per year, and interest rates at 5%, you have a budget of about $600 to work with to break even. Remember, the key is to stop the sunlight from reaching the window glass in order to achieve the maximum benefit.

By the same token, if you have the option, locate the outdoor portion of your central air conditioning unit so that it is in the shade in the afternoon. That reduces the amount of work it has to do in ejecting the heat from your house. Water-cooled condensers should reduce energy use even further.

If you think installing photovoltaic (solar) panels to produce the electricity is the solution, consider these numbers. To overcome the solar gain effect, you need to produce 1.5 kW of power—1500 watts—while the sun is shining. Assuming an installed cost of Cdn $15.00 per peak watt for solar panels (Cdn $7.50 per peak watt for the panels and an equivalent amount for mounting hardware, wiring, interconnection equipment, labour etc.), that comes to Cdn $22,500.00. Sixty hot summer days per year and a requirement of 10.5 kWh per day for air conditioning to overcome the solar gain heating, it would take over 200 years for these solar panels to pay off (and they only last about 25 to 40 years producing their rated output). Looking at it another way, the $22,500 investment will save you less than $2,500 over their lifespan—a net loss of $20,000. You will actually obtain a greater benefit from the solar panels if you hang them as awnings over the sun-facing windows to block the solar gain than you will from the power they generate.

## Use Natural Ventilation and Cooling

During the cooling season, when temperatures are cooler (e.g., at night), open windows to cool your building with cooler air, and give your air conditioner a break. If the air is still, use fans to move the air. Where practical, draw in cooler air down low, and exhaust warmer air from a higher opening (e.g., an attic exhaust fan), to take advantage of convection effect. Take care not to sacrifice your security by leaving unsecured basement and ground-floor windows open when you are asleep.

If you have a basement, it is probably cooler than the upper floors. The ground around your basement and below the frost line will typically stay close to 12°–15°C (55°–60°F) year-round in southern Canada and the northern U.S. Take advantage of that natural coolness when it is practical.

## Use Fans

Use ceiling fans, oscillating fans, whole house exhaust fans, or other air-movers instead of air conditioning where practical. Fans use much less power for similar perceived cooling effects than air conditioners.

In general, when ceiling fans are used for cooling, it is preferable to have them pulling air up, instead of driving the warmer air that should be near the ceiling down. However, this may not work in all circumstances.

## Reduce Unnecessary Heat

Switch to energy efficient lighting—efficient lights produce less waste heat.

Cook less. Serve cold meals on occasion, or use the barbecue. Use the microwave instead of the oven or stove where possible.

## Solar Chillers

If you have a large facility to cool on hot sunny days, determine whether a solar-powered absorption or adsorption chiller might be an effective alternative.

An absorption chiller uses heat to drive the refrigeration cycle instead of a mechanical compressor as is commonly used in conventional refrigerators and air conditioners. Absorption chillers are used in propane-powered refrigerators and in many indoor ice rinks. Adsorption is the new kid on the block. It can use lower temperature input and produce lower temperature output, uses silica gel instead of lithium-bromide, and is expected to have a longer working life. Both absorption and adsorption chillers can use waste heat, or solar energy, or both, to provide the input energy.

Using solar energy for air cooling (using an adsorption chiller) has the benefit of matching energy supply—hot sunny days—to demand. So far, absorption and adsorption chillers are sufficiently expensive and large that they are not appropriate for residential applications.

## 26.4. Hot Water

After climate control, domestic hot water is our next major household energy consumer. I'm certainly not advocating that we learn to live without it. I happen to believe that hot running water is one of the real marks of our civilization, and a hot shower is one of those daily conveniences I truly appreciate. However, there are things we can do to make that use more efficient and sustainable, and reduce the impact on our finances.

These tips will help you reduce your energy bill, whichever fuel you use to heat water.

### Use Less

Switch to cold water for all your laundry. There are detergents that will work well on all clothes in cold water, including whites. Try to run with full loads in automatic washers to reduce the number of loads done. If washing smaller loads, remember to set the water level accordingly.

Wash your hands in room temperature water, from the cold water tap. You move a lot of hot water into the hot water pipes just to get a couple of seconds worth at the sink. How long do you wait for hot water to arrive? 30 seconds? 60 seconds? Well, hot water is leaving the hot water tank the whole time you are waiting. And cold water is replacing that water in the tank, which you will pay to heat. Are you using 60 seconds worth of hot water from the tank to get 5 seconds worth for your hands? Typically, the first few seconds of water from the cold-water tap is at room temperature—it has warmed up simply by sitting in the pipes inside your house. Water at room temperature with soap is adequate for washing your hands most of the time. Using the first few seconds worth of water, instead of waiting for the hot water to arrive saves time too.

Water saver faucets will reduce your hot water consumption, by reducing your water consumption in general (covered later in the book).

### Insulate the Hot Water Tank

Increase the insulation around your hot water tank. There is a little built into the tank itself, typically about 3 centimetres (about an inch) of fibreglass between the actual tank and the outside metal skin. It isn't enough (manufacturers limit it to keep costs down and ease transportation of the tanks). Most hot water tanks are installed in the coldest part of the house, in the basement, near a cold, concrete, exterior wall. Insulating jacket kits are available at most hardware stores. Or you can make your own jacket or enclosure. If your water heater is fired by natural gas, make sure you do not block the air intake or exhaust. Remember, your gas water heater is using fire to heat the water. Put more insulation on top of the heater, and between the heater and the nearest wall to maximize results. Keeping the heat in the hot water tank will help keep the air around it cooler, which may help with that air conditioning bill as well.

## Insulate the Hot Water Pipes

This is of most benefit if there are short periods between hot water uses, e.g., people showering after each other. The insulation helps the water in the pipes stay warmer, longer. It also reduces the amount of heat radiated away from the pipes as the hot water is flowing, so it should arrive at the destination slightly hotter. That may permit lowering the temperature setting on the water heater thermostat slightly.

## Set the Temperature Lower

Turn down the thermostat setting on the hot water tank to the lowest setting you find tolerable. This reduces the amount of energy used to raise the water temperature and maintain the desired temperature. Many people are satisfied with hot water at 50° to 52°C (120° to 125°F). The water heater tank thermostats are often set at the factory or by service people to 55° or 60°C (130° to 140°F). Use of an automatic dishwasher may limit how low you should set the hot water control—again 50°C (120°F) is typical.

Any temperature much above 52°C (125°F) at the tap creates a risk of scalding from tap water. Temperatures below 50°C (120°F) are not recommended, as some hazardous bacteria can survive in warm water less than this temperature (e.g., legionellae).

## Point of Use Water Heaters

If you have long runs of hot water pipe, or use hot water infrequently, you may wish to consider point of use water heaters, sometimes referred to as tankless water heaters. These are typically heated with electricity or natural gas. The electric units may include a small reservoir to provide instant hot water, and continue heating water as it flows to the tap. A variation on this is to install a second conventional (if smaller) hot water heater in the remote location, which also eliminates long runs of hot water pipe.

These are not a no-brainer. Seriously consider all the costs associated with an installation (e.g., additional wiring and gas-pipe runs) vs. the expected savings before making the decision to purchase.

## Boiling Water

This is the pinnacle of applying energy to our domestic water. Typically, it is very small quantities, used for hot drinks (e.g., coffee, tea) or in cooking. The small quantities of energy used in this application typically do not justify major investments in sustainable energy devices, but even here there can be efficiency gains. For example, we can boil water in an open pot on a stove, typical for cooking some foods. One way to increase efficiency is to put a lid on the pot. However, if our objective is to boil a cup of water to make tea or coffee, a small electric kettle will be more efficient. On the other hand, if making a cup of hot beverage is something we do very rarely, it may not be worth the investment in an electric kettle. These are the kinds of factors we need to keep in mind for many of our energy-related decisions. Whether or not a proposed change makes sense always depends on the context.

An electric kettle uses less power to boil water than a stovetop kettle. This is because the heating element is in direct contact with the water in the electric kettle. With the stovetop kettle, only about 25% of the heating element is in contact with the bottom of the kettle, leaving the rest of the element to heat the stovetop and surrounding air, and the kettle via convection.

An electric kettle is also more efficient than the microwave for boiling water.

Boil only as much water as you need at a time. Heating more wastes power.

Makers of "boiling" hot water dispensers claim their devices use less power than even microwaves or electric kettles for providing boiling water. The water actually comes out at about 88°C or 190°F for safety reasons. I expect this is true where a significant number of cups of water are being served, as the dispenser delivers precisely the amount of water desired, so no extra water is being heated and then allowed to cool again, and the water is not heated to as high a temperature.

## Pool Heating

Look into buying a solar blanket for your pool. This will heat the water when the sun is shining and provide insulation to retain heat when the sun is not co-operating.

## *26.5. Appliances*

### Refrigerators

Much progress has been made in energy efficiency for mainstream refrigerators in the past fifteen years. The industry as a whole has embraced more efficient construction and components. A current model of 18 cubic feet consumes less than one-third of the energy of a comparable unit manufactured fifteen years ago or more. It is a little-known success story in the energy-efficiency business. If you make the effort to compare Energuide or EnergyStar labels while appliance shopping, it can really pay off.

If you use a conventional electric refrigerator built before 1994, replace it. Check the energy consumption numbers, shop around for a good deal, and make the purchase.

Here's what we found. Our old refrigerator, built in 1989, was rated at 97 kWh per month for power consumption, or 1164 kWh per year. It wasn't a glutton, just typical for its time. Although it was working well, we wanted to replace it for purely financial reasons. The new fridge, not the cheapest or most efficient available, consumes 374 kWh per year. We save 790 kWh per year. At $0.15 per kWh, our annual direct savings are $118.50. However, there are additional savings because there are 2 kWh a day of heat not being produced in our house during the cooling season. We estimate these savings at approximately $20.00 annually. We also now have a refrigerator with a 10-year warranty; there was no warranty remaining on our old fridge. Not as good a "return on investment story" as the compact fluorescent lights, but still a positive return. With careful shopping and a more efficient refrigerator, you could do better.

Imagine if all our appliances could reduce their energy consumption by almost 70% in a decade, with no sacrifice in features or utility! It has happened in residential lighting (with energy-efficient lighting options) and domestic refrigeration.

Do you have a second fridge that you only use part of the year? See if you can compress all the items you refrigerate into one fridge. Unplug the second (older) fridge when it's not needed. If it is more than ten years old, it is almost certainly an energy hog compared to modern refrigerators. If it consumes 3 kWh/day (typical), your savings will be $13.50 per month. Make sure you prop the door open to permit air circulation to stop mould from forming while it is unplugged.

Pull it out from the wall and, if there are coils at the back, vacuum them. The accumulated dust on the coils acts as insulation, inhibiting heat exchange, making the compressor work harder and draw more power.

Keep the fridge fairly full to minimize air space. Use bottles of water if necessary to reduce air space.

Keep the fridge door closed as much as possible.

Make sure the temperature control is set to the correct temperature (many are set too cold, consuming more power).

There is increasing interest in alternatives to conventional refrigerators for preserving food (e.g., Peltier devices, absorption or adsorption chillers, modern variants on the ice-house), but nothing yet that is ready for the mass market. I'll be keeping my eyes open.

Here are my favourite refrigeration energy efficiency stories. I conducted a couple of experiments on an older chest freezer and an older refrigerator.

The chest freezer is a 30-year-old, conventional unit, similar to many still in service. This one has a capacity of approximately 7 cubic feet, and consumes 105 kWh per month. We placed an extruded polystyrene board under the freezer to block convection heating from the floor into the freezer. The insulation board was approximately 4 cm (1.5 inches) thick with an R-value rating of 7.5. The board was cut to be 2.5 cm (1 inch) larger all around than the footprint of the freezer. This simple measure reduced the energy consumption of the freezer by 27%, to 77 kWh per month. Subsequently, we attached another piece of polystyrene board to the top of the freezer. This addition further reduced the energy consumption by 7%, to 69 kWh per month. Total savings: 34%.

As this type of freezer uses the side walls to house the condenser coils, insulating the sides would have been counter-productive.

The experiment with the refrigerator began with an older refrigerator, with condensing (heat) coils on the back of the refrigerator. Again, it is similar to the older, second refrigerators found in many homes. The thought behind this experiment was that the heating coils direct 30% to 40% of their waste heat back towards the refrigerator. The modification was to spread the condenser coils as far away from the refrigerator as possible, and to insert a layer of reflective foil and insulation (in the form of aluminized bubble-wrap) between the refrigerator and the coils. The result of this simple addition was to reduce the energy consumption of the older refrigerator by approximately 25%. Not as great a gain as buying a new, energy efficient fridge (up to a 70% savings on energy), but a better return on the investment.

And finally, in 2004 we bought a new refrigerator, primarily to get the energy savings it promised for our electrical bill. I watched for these savings on our bill, but only about half of what I expected materialized in terms of savings. Some time later, I had the opportunity to use an electricity consumption meter on this refrigerator. Once it was in place, I noted that the instantaneous consumption was 38 watts. I felt this was odd, given the compressor was not running, nor anything else as far as I could tell. I opened the refrigerator door to see whether I could locate the source of the power draw. The meter still read 38 watts. That's when the light in my head went on—the light in the refrigerator had never gone off. Whether the door was open or closed, the consumption was 38 watts. I pressed on the door switch by hand, and the energy consumption went to zero. The door is properly aligned from the outside, but the bump on the door interior is just low enough so that it does not press on the switch. I taped a piece of cardboard to the bump, and the problem was solved. The savings amount to about a kWh per day, every day. If electricity costs $0.15/kWh retail (including all taxes and surcharges), that comes out to about $55 a year, for a small piece of cardboard and tape. Moral of the story: use an electricity consumption monitor to test your appliances to see where you can save.

## Washing and Drying Laundry
### Washing

This is an area where I expect to see great strides in energy efficiency over the next few years. As innovations from other markets and different technologies are adopted (e.g., horizontal axis machines, water saver designs), both energy and water consumption will be reduced in the washing machines of the future relative to what we use today. Some will probably even use less detergent to clean the clothes.

If you expect to be replacing a washing machine in the near future, I recommend looking into a wide variety of options starting as early as possible. Check the EnerGuide or EnergyStar numbers, and the estimated water consumption. Check all the features to get a unit that suits your needs (e.g., rinse water re-use, variable load levels, variety of wash cycles). Consider all the costs (purchase, power, water, detergent, size of loads, your needs) before making your decision.

That said, don't forget to use your existing machine to its maximum benefit. Use full loads when possible. Otherwise, use the variable load feature if available. Re-use rinse water where feasible. Energy and water saving features are frequently provided in the owner's manuals for washing machines. Take the time to read them.

## Drying

To reduce your energy costs, you can use a solar and wind power clothes dryer, also known as a clothesline. For those not able to hang laundry outdoors (no facility, local bylaws, extended rainy periods, etc.), indoor clotheslines are viable.

Retractable lines (typically up to 6 metres or 20 feet in length) are commercially available at reasonable prices. Retractable clotheslines allow the space used for hanging clothes to be used for other purposes at other times. Mount both ends securely on solid surfaces. Wet laundry can be heavy, and the weight will easily pull screws out of drywall—that's the voice of experience speaking.

A natural gas clothes dryer uses electricity as well as gas; the motor that turns the drum uses electricity, as do the timer, buzzer, and other electronics.

An electric clothes dryer consumes a lot of power when it is running, typically more than 4 kilowatts. When you do use your clothes dryer, make sure you have full loads. Avoiding an extra cycle is the easiest way to reduce the amount of electricity used.

Empty the lint trap before every load. This reduces the amount of work the blower has to do to move the moisture out of the clothes.

Periodically, wash your lint trap. It seems that fabric softeners will adhere to the mesh of the lint trap over time, making an almost invisible film, blocking the airflow. Washing the lint screen removes the film and improves the airflow, so your clothes can dry faster and use less electricity.

When shopping for a clothes dryer, look for a smart unit that senses when the clothes are dry, rather than relying on a timer.

Set the dryer to stop before clothes are completely dry. Remove permanent press items while still hot, and hang them or fold them. In many cases, this eliminates the need for ironing.

Make sure there is a damper door on the outside of your dryer vent so that cold air is blocked from blowing in when the dryer is not venting to the outdoors.

## Clothes Irons

Only plug the iron in when you need it. When it is on, do your ironing in a concentrated session, and then unplug the iron. It uses electricity the whole time it is plugged in, whether you are using it or not.

# 26.6. Dishwashing

Hand washing can use less water and energy than a dishwasher, but not if the hot water is left running constantly (e.g., to rinse each dish before putting it in the drain-rack).

Letting small quantities of dirty dishes soak in a dish tub makes the dishes easier to clean later, and means fewer sinks worth of hot water in total than washing small quantities at every opportunity. Wash the dishes in moderately sized sessions. The dish tub itself makes a reasonable indicator of how many dishes to save up for a session. Marathons at the sink are not much fun. Reasonable sessions can be an opportunity for family members to talk.

Don't run the dishwasher until you have a full load. Reducing the number of loads is the easiest way to reduce the amount of power used.

Don't use the hot-dry cycle—let the dishes dry at room temperature. The hot-dry cycle uses the electric heating element.

Use the most economical setting that will do the job. Many dishwashers come with heavy-duty and normal settings, and some with rinse or economy settings. Use the one that uses the least power (typically shortest operating time) but still gets your dishes clean. Letting dishes soak in a dish tub for a few hours before putting them in the dishwasher may let you use a lower-power setting.

# 26.7. Cooking and Food Preparation

## Food Preservation and Preparation

Why not just say cooking, refrigeration, and freezing? Because there are many other alternatives that we preclude by narrowing the topic. We need the sustenance provided by food and drink to survive, and we need to protect ourselves from hazards that accompany our food by taking appropriate safeguards against various bacteria or moulds that can make us sick.

## Small Appliances

In the course of food preparation, you can use hand-powered devices like whisks, knives, can-openers, and choppers rather than electric appliances like blenders, food processors, mixers, and can-openers, especially for small quantities.

## Stoves

Instead of using a full-size range-top or oven, look for other options for heating food when feasible.

An electric kettle uses less electricity to boil water than a stovetop kettle.

A microwave generally uses much less electricity to heat food than a stovetop burner or oven. A toaster-oven uses less power to heat small quantities of food than a conventional oven.

When you are cooking on a stovetop burner, cover pots and pans as much as possible to keep the heat in, rather than heating the kitchen.

Let food defrost in the refrigerator overnight, rather than cooking from frozen, to reduce energy consumption.

When using the oven, resist the temptation to open the door and peek in on the food as much as possible to help keep heat in. If the oven door has a window, use the oven light instead.

Most conventional ovens have a heat exhaust that comes up under one of the rear burners (not the case with newer flat-surface stove tops with ceramic surfaces). Make use of that heat when possible, e.g., to warm water for cooking vegetables.

Keep the bowls under your burners clean, so they will reflect heat back to the pot. Better still, use reflective aluminum liners, and keep these clean or replace them when they are dirty or dulled from age.

On ceramic top stoves, allow for the fact that the burners take a long time to cool by turning them off a couple of minutes before you are finished cooking on that burner. The residual heat will finish the job.

In summer, consider cooking outdoors (barbecue) instead of heating the building with the stove. There's a reason our forebears had summer kitchens in many homes.

## Barbecues

The barbecue is the focus of that marvelous social event, the barbecue party. Grilling foods as a cooking technique can help reduce the fat content of meats by draining it. In most of North America today, a barbecue means propane.

There are other options, including natural gas and electric barbecues. If you are using your natural gas barbecue effectively, it is not consuming much fuel, and the economics of switching to another fuel are questionable.

Electric barbecues are usually limited to 1500 watts (by our typical household electrical circuits). These simply don't make the grade when you want to cook twenty steaks or fifty burgers in short order. They also don't appeal to the bon-fire-style barbecuer. However, they are superb for smaller family meals, or more complex cooking tasks where temperature regulation is more important. No emissions. No tanks to refill. Power outages could be a problem.

One option that may be worth considering is solar cooking. Typically more appropriate for foods that require lower temperatures (100° to 150°C or 200° to 300°F), this environmentally friendly means of cooking seems to be catching on in North America in ever-increasing numbers. Not always a direct substitute for the barbecue, the solar oven or box cooker is more like a slow cooker than a stove, to use kitchen appliance analogies. However, parabolic solar cookers can deliver a lot of heat in a short period of time. Solar cookers will get more coverage later in the book.

On really hot days, consider forgoing the barbecue for cold meals. Your body and the local environment don't need the extra heat on a day that is already very hot. And the air quality will benefit as well.

## 26.8. Other Appliances and Electrical Equipment

### Turn it Off

The biggest area for electrical savings for most of us is still the off-switch. Turn off lights, appliances, and equipment when they are not required.

### Ghostbusters

Many devices consume power even when they are off. These are known as phantom loads. For example, most televisions consume power to support their instant-on features. So, even when you think they're turned off, they're not. The only way to stop these devices from consuming power completely is to isolate them from the wall socket, either by unplugging them or using a discrete isolation switch, as may be found on power-bars, surge-protectors, or purpose built switches. If that has not spooked you, consider this: Lawrence Berkeley National Laboratory (U.S.) says electrical appliances that were switched off consumed over 71 terawatt-hours in the U.S. in 1999. That's 71,000,000,000,000 Watt-hours—to power appliances that are off. Enough power to light four 15-watt compact fluorescent lamps for 4 hours a day for more than 810,000,000 homes for a year. Still too big to comprehend? It's about *twice* the power that the entire Pickering power generation complex could produce if it could run all *eight* of its reactors full-out for a year. Yes, the equivalent of about sixteen large reactors running at full capacity, continuously.

Constant-operation battery chargers (also known as "wall-warts", those little black boxes that plug into the outlet and then feed out a wire to the device, often with a spear connector) are another type of phantom load. One novel solution to this electrical load is a solar charger that will charge the batteries daily, without using any electricity from the power grid.

Another possibility is to attach your wall-wart chargers to a power-bar, and put the power-bar on a timer. This way, all the devices draw power for only a few hours each day, which should be enough to maintain batteries. An even better idea if you can benefit from interval pricing.

## Computers

The largest power consumer on your computer is likely the monitor. Shut it off when it is not in use. This is not the same as the screen saver, which may just generate a dark image, but not actually save any power. If your monitor has a time-out power-saver feature, enable it. If it has an adjustable time setting, set it to the shortest time that works for you, perhaps 2 minutes.

The newer flat LCD monitors typically consume less energy than the conventional cathode ray tube type, and are becoming increasingly affordable.

Laptops typically consume less power than desktops or towers, to maximize life on a battery charge.

Other big power consumers are your printers and other peripherals (e.g., scanners). Shut them off when not in use, even for short periods. When shopping for printers, look for those with power-saver modes, or auto-shut off based on timers.

## Appliance Shopping

When it is time to replace an appliance, shop for the one that uses less electricity. For example, smaller refrigerators usually use less power than larger ones. However, insulation and design make a difference, so look for the EnerGuide or EnergyStar label to get a common benchmark for comparison across models. For smaller appliances without EnerGuide or EnergyStar labels, check the plate or power supply label for power ratings (in watts or amps—lower is better). For hyper-efficient devices, look at what people living off-grid use for appliances. Some jurisdictions offer tax rebates or other incentives on the purchase of efficient appliances. Appliance retailers will typically know what incentives are available. Otherwise, check with appropriate levels of government in your area. If rebates are a determining factor in your decision, make sure you comply with all the fine print before buying.

### Time of Use

If you can time your major loads for off-peak times, this will reduce the peak prices your utility has to pay for peak-demand (expensive) power. For example, run your dishwasher or electric clothes dryer after 7 PM, when demand for electricity is lower. After 8 PM is even better. The lowest demand period during the day is typically from 3 AM to 5 AM.

This will be even more attractive for those that qualify for time-of-use pricing, also referred to as interval pricing or smart metering.

## *26.9. Smart Meters Need Smarter People*

Ontario is embracing smart meters as one means of dealing with the current situation in which generation capability does not meet peak demand on a regular basis. This situation will be compounded by plans to eliminate over 20% of current generating capacity by shutting down all coal-fired production. A lot is riding on the ability of Ontarians to shift their usage patterns to make this work.

A smart meter isn't really so smart. It is a recording device that can tell when you used a specific amount of electricity and report it, either to your utility, to you, or both. With adequate communications capability, this information can go to the utility on a daily basis, or to your home PC even more frequently.

The catch to the introduction of smart meters is the parallel introduction of tiered pricing of electricity depending on the time of day it is used. While the pricing structure has not been finalized, early predictions suggest that the commodity price of electricity will be about three times as high at regular peak demand periods (e.g., 4:00 PM, Monday to Friday) vs. the lowest demand periods (e.g., 4:00 AM everyday). This reflects the price of peak power from generators vs. base supply. This is going to end up costing the average consumer more money, despite anything the politicians may have to say about the measures being revenue-neutral or the average price over the course of a day, week, or month remaining the same. That's because these semantic exercises ignore the reality of what a peak demand period is.

Let's take a couple of simple examples to try to demonstrate this. Suppose the current market prices electricity at 10 cents per kWh, and our ratepayer uses 2 kW on average. As the rate is the same regardless of when the power is used, our ratepayer will pay $4.80 per day for their 48 kWh of electricity use (24 hours x 2 kW = 48 kWh; 48 kWh x $0.10 = $4.80).

Next, let's move to a scenario where electricity costs 5 cents a kWh from 7 PM to 7 AM each day (off-peak period) and 15 cents a kWh from 7 AM to 7 PM. The average price is still 10 cents per hour ((12 off-peak hours x $0.05 + 12 peak hours x $0.15) / 24 total hours per day = $0.10). If our ratepayer uses electricity at a constant 2 kW, then the total cost will not change.

However, if they are like most people, they will use more electricity during peak periods (that's what makes it the peak demand period) than during the off-peak period. Let's assume they use 1 kW during the off-peak period and 3 kW during the peak period. Their daily consumption has not changed; it is still 48 kWh per day (1 kWh x 12 hours + 3 kWh x 12 hours = 48 kWh). But their cost has jumped dramatically, from $4.80 per day to $6.00 (12 off-peak kWh x $0.05 + 36 peak kWh x $0.15 = $6.00). No change in overall consumption, but a 25% rise in their bill!

In reality, it could be worse for residential electricity consumers, depending on their current electrical use patterns and the real differential between peak and off-peak prices.

However, there is a potential silver lining in this. Cheap electricity. That's right, at the lowest demand times, electricity will actually cost less than it does under the current single-price system. It is up to price-conscious consumers to figure out how they can benefit from the cheaper electricity available at night, on weekends, and holidays.

Here are some suggestions, though this list will be far from exhaustive.

Run your electric clothes dryer at cheap power periods. Your clothes dryer can use 4 kWh or more per load. It's your decision whether that will cost you $0.20 or $0.60 (based on our example figures above)—three times as much. If your household does five loads of laundry a week, the potential price difference could be $2.00, or over a $100 a year.

Same deal with your dishwasher. It is not as power-hungry as your clothes dryer, but it won't make any difference whether your dishes were washed at 9 PM instead of 11 PM.

This may also apply to major use of your electric oven. If possible, try to do your major oven cooking operations on weekends and holidays, and not during weekday electricity rush hour.

Timers can also be advantageous in this situation. If you have a number of devices that draw power continuously but don't really need to during peak periods, a timer can solve the problem. One example is batteries for power tools that sit in chargers in preparation for their next use. Put the chargers on timers so they only charge at off-peak times, and not during peak periods. The batteries will still be recharged daily, but using cheaper electricity.

Use the same approach for automotive block heaters.

Programmable thermostats can accomplish the same thing for electric heating systems and central air conditioners. For window air conditioners, use a heavy-duty, discrete timer unit.

If you have a chest freezer, consider this approach. Put the freezer on a timer so that it only draws power at off-peak times. Get some 2-litre bottles (used plastic soft-drink bottles), about one for each 2-3 cubic feet of freezer capacity. Put 150 millilitres of table salt in each bottle, and then add water until the bottle is about 90% full. Cap tightly. Put these in the freezer as thermal mass to stabilize the temperature. The brine should freeze at about -10ºC. Your freezer should have no difficulty freezing the brine. However, when the freezer is without power, the brine bottles will stabilize the temperature at approximately their freezing temperature for many hours—longer than required for the daily power cycle. If possible, place the bottles in a basket that suspends from the top of the freezer, so the bottles are closer to the top of the compartment.

This approach can also work in a refrigerator freezer compartment, if it is not opened too often during the peak price period. However, refrigerators usually see more use during peak-price periods, so it may be more appropriate to limit their controlled off-time to avoid only high-peak periods (vs. mid-peak periods in those regimes with three-tier pricing).

If you need to replace an electric water heater, look for a unit that includes a timer or permits the use of one. These have become less common in recent years for 240-volt devices, but may become more popular again with the tiered pricing system. Most hot water heaters have plenty of reserve to deal with the evening meal without additional power. Let the water tank reheat when electricity is cheap. It will be ready for your morning wake-up shower.

Let's revisit our example above, but turn it around a bit. Suppose you manage to move the bulk of your power consumption to off-peak periods using the techniques above, and similar approaches. This won't be easy for most of us; it will take some serious effort to make changes in our behaviour. Imagine that our rate-payer moves about 55% of their power use to off-peak periods (majority of use of the clothes dryer, dishwasher, heavy cooking, freezer, block heater, air conditioner), instead of the 50% she or he needs to keep costs unchanged under two-tier pricing. Now, the electrical bill will be $4.50 (27 off-peak kWh x $0.05 + 21 peak kWh x $0.15 = $4.50), lower than under the single rate system.

## 26.10. Lighting

### Energy-Efficient Lighting

The conventional incandescent light bulb is horribly inefficient when it comes to producing light. It produces more than 3 times as much heat as light from the power it consumes. That's a lot of heat. It is actually sufficient to cook food, as was demonstrated by a toy oven in years gone by, which used a single incandescent light bulb as the only heating element.

Fluorescent lighting, including compact fluorescent lights (CFL), consume about 25% of the power of incandescent bulbs to produce the same amount of light. And fluorescent lights last longer. Compact fluorescent lights will usually fit right into your existing incandescent light fixtures without modifications. In some cases, CFLs will not fit into existing shades or light covers. Switching to fluorescent lights will save you money in two ways. They will outlast incandescent bulbs, reducing purchase price and maintenance effort over their lifetime. They will also reduce your electrical consumption by about 75%. The variety of CFL lighting continues to expand. There are now flood and spotlight packages, and more types rated for outdoor use—even coloured lights. Warmer colours have improved the appearance of fluorescent lighting in the home environment.

One great application for CFLs is work lights, sometimes called trouble lights. They don't get as hot, and stand up to vibration and movement much better than incandescents.

A more recent, and currently more expensive, contender for efficient lighting is Light-Emitting Diode (LED) lighting. Research into better white LEDs continues, but expect to see more of these in the future when energy consumption and price both come down. LED lighting tends to be very directional, which is more appropriate for task lighting than room lighting.

Coloured LED lighting is already commercially available, with attractive life-cycle cost relative to incandescent mini-lights for decorative lighting. Life expectancy for these lights is in the hundreds of thousands of hours.

### Natural Lighting

It's free! Arrange furniture to take advantage of natural lighting. During the cooling season, you may wish to keep windows covered to reduce solar gain (heating from the sun). Even so, unless you have blackout curtains or completely opaque shutters, plenty of natural light will likely leak in.

## Decorative Lighting

Cut back. Put up fewer lights. Reduce the number of hours they are lit over the course of the season. Use decorations that are more visible in existing light. Look for the new LED based decorative lights—they consume a lot less power than the traditional incandescent bulbs, as little as 10% of conventional decorative lighting. We are very pleased with our LED decorative lights, both indoors and out.

## Sensors and Timers

Put motion sensors, daylight sensors (photosensors) or timers on outdoor lights to reduce the amount of time they are on. Daylight sensors are also available for some small nightlights.

Consider the case where there is a 100-watt light left on all the time to provide outdoor lighting. In a year, it will consume 57.6 kWh. Suppose we replace this with a 25-watt compact fluorescent light with a photocell sensor. On average, this light will now be on less than 12 hours a day, only when it is needed to provide lighting. Assuming it is on 12 hours a day average over the year, this light will consume just 1/8 the power of its predecessor (7.2 kWh), while providing the same amount of light during dark hours. That's a saving of 50.4 kWh, over 85% less energy consumed.

If electricity costs $0.15 a kWh, that's $7.56 per lamp per year. If the incandescent bulbs last an average of 1,000 hours each (optimistic in my experience), that is 5 bulbs saved, likely worth more than $5.00 in the year, plus the effort to change the bulb 5 times. Let's round that down to a $12 / year annual saving. The costs are about $5 initially for the photocell adaptor socket, and a replacement fluorescent bulb approximately every two years at a cost of about $7 (still falling). Payback period? 7 months. After that you are saving money forever. Return on investment? At least 630%. Another advantage for photocell sensors is that they will turn off lighting when power is more expensive on summer afternoons, when there is plenty of light available.

## More Switches

For areas where you have a lot of lights on a single switch, as you may in your basement, you may find that larger areas than necessary are lit. You can install switches at some of the fixtures so that they can be controlled individually, and reduce the number of lights that are on.

### Perspective

Atikokan is a coal-fired generating station in Ontario. It produces up to 230 megawatts of electricity. That's about 2 kWh per day per Ontario household. If each household replaced three 100-watt incandescent lights that are lit an average of 8 hours a day with compact fluorescent lights, not only would they save about $100 on their annual electrical bill for their $15 investment, we could remove the equivalent of the Atikokan coal-fired plant running full-time from the demand on the Ontario electrical grid.

Actually, Atikokan is one of the newer, cleaner coal-fired plants. From an environmental perspective, we would be better off leaving Atikokan in operation, and shutting down an equivalent amount of power generation from an older, more polluting coal-fired generating station.

## 26.11. Miscellaneous

Use windup clocks instead of electric clocks.

For cold-weather car starting, a battery blanket uses less power than a block heater. If you are going to use a block heater anyway, use a timer so that it is drawing power for only an hour or so before you need to start the vehicle, instead of all night.

## 26.12. Water Consumption

Most of us don't think of water as part of our energy consumption, or even of it being transported, because it is always available at the taps in our homes and other facilities. In reality, for most of us, the water we use travels many kilometres (miles) through a labyrinth of pipes, pumps, reservoirs, and treatment facilities before appearing at our local faucet.

Another reality is that water is heavy, so it takes a lot of energy to move it over long distances, and most often, uphill. If you don't think so, note that sewage, the inevitable return leg of the round trip, almost always travels downhill, powered primarily by gravity. The cost of that energy shows up in the form of our local utility bills, or tax bills.

These water bills are primarily about transport energy. The utility doesn't make the water. They just treat it and move it.

If you think this is a trivial amount of energy, try carrying the water you use in your household each day for a few kilometres, especially uphill. Direct household use in North America is typically in the order of 300 litres per person per day. Direct overall use by Canadians averaged 343 litres per person per day in 1998.[1] Municipal figures show a usage almost double that, which includes uses such as firefighting, street cleaning, and filling pools. For example, in 1999 we used 638 litres per person per day.[2] U.S. figures[3] indicate that per capita water use is similar, in the order of 650 litres (185 gallons) per day (excluding agricultural and irrigation usage).

I have figures for an Ontario city of approximately 120,000 in population. In 2000, the total electrical consumption for the municipal government was approximately 30 gigawatt-hours. Approximately 45% of that electrical consumption (13.5 gigawatt-hours) was associated with municipal water service; pumping, reservoir operation, and sewage treatment. That's about 115 kWh per resident per year. By way of illustration, that's enough electricity to keep a conventional 13-watt compact fluorescent light (60-watt incandescent equivalent) lit continuously (24 hours a day, 365 days a year) for each of the 120,000 residents of that city.

For those of us on wells or cisterns, the water may not travel as far, or be aggressively treated, but we still use energy (i.e., pumps) to move it and pressurize our household plumbing systems, and we see this cost on our power bills.

Obviously, the simplest way to reduce our energy expenditures related to water consumption is to use less water. There are many ways to go about this. One way is to get multiple uses from it. A triple-header from our household is to use water to boil eggs (or cook vegetables), then let the water sit for a few hours to extract the heat from it, and finally use it to water the houseplants.

Some tips on reducing energy consumption associated with hot water were provided earlier in the book. The following tips are about reducing water consumption in general.

## Toilets

In most North American households, this is the number one consumer of water, typically accounting for about a quarter of residential water use. It is a prime place to look for some savings.

Toilets actually consume more water per flush than the nominal volume of the toilet tank. This is because some water is also supplied to the bowl in each flush, via the hose from the fill tower to the overflow pipe while the tank is filling. Some

---

1.    http://atlas.gc.ca/site/english/maps/freshewater/consumption/domestic/1
2.    Ibid
3.    http://www.pepps.fsu.edu/safe/pdf/sc1.pdf

228    The Emperor's New Hydrogen Economy

of this actually ends up flowing out of the bowl and into the wastewater system. For a 6-litre (1.6 gallon) toilet, this additional consumption averages about another 3 litres (0.8 gallons), or another 50%! For the older 16 litre (3.5 gallon) types, this additional consumption is approximately 4.75 litres (1.25 gallons), about 30%. So, if you couldn't figure out why your water bill showed you using so much more water than you thought you were using, this could be the explanation.

There are after-market devices that divert some of the water destined for the over-flow tube back into the tank, reducing water consumption. I have used one such device. It works as advertised, without issues. I have not seen these in stores, as yet. You can see the one we use at:

http://www.savewaterandmoney.com

Older, conventional toilets use about 18 litres (4.75 gallons) per flush. 5 flushes per day per person is typical. (Yes, there really are studies, which show that people flush anywhere from 4-7 times per day. My guess was about six from our household, but we'll use five as it seems to be the median figure from the studies I have found.) If there are four occupants in the house, that's a total of twenty flushes. If we go from an older conventional toilet (18 litres per flush) to a newer low-flow toilet (9 litres per flush), the total savings comes to 180 litres per day (20 flushes x 9 litres per flush savings). Expanding that to a year, that's 65,700 litres. That's a lot of water. Think swimming pools! Here in Ottawa (in 2006), where we pay about $1.75 per cubic metre of water, that's an annual savings of more than $115. This savings should more than pay for the original low-flow toilet. The savings after that go right into your pocket.

Assuming you are on a municipal water supply, there should also be tax savings, as there will be less need for building more infrastructure due to reduced demand for water and wastewater flow.

In addition, less energy is consumed moving water. Sorry, this is going to take some physics and math. Let's suppose you live in a city in a high-rise apartment, on the 15th floor of a building with a nice view, located 15 kilometres from the local water treatment plant, and 300 metres above the water source for the treatment plant. The water has to travel 15 kilometres horizontally, and 350 metres up to this apartment. We'll even ignore the energy used to treat the water in the plant, and the energy used to push it through the pipes on the nominal horizontal, and focus just on the 350 metre lift. A litre of water weighs one kilogram. It takes ten joules to lift one kilogram one metre (overcoming the force of gravity at the earth's surface, specifically sea-level). So, it's going to take (350 metres x 65,700 kilograms x 10 joules/metre-kilogram) 229,950,000 joules to lift the water we saved in a year. That's an impressively large number! However, let's convert that to something more familiar. A joule is a watt-second. So, a kWh is (1000 watts x 3600 seconds in an hour) 3,600,000 joules.

When we do the division (229,950,000 / 3,600,000), we end up with about 64 kWh saved. Per household. Per year.

There are also toilets that use even less water, but these tend to be considerably more expensive, and are not acceptable in all jurisdictions. There are also zero-water composting toilets available (e.g., Envirolet), but these are not acceptable in many jurisdictions, including most urban areas.

## Texas Water Rules

This is not intended as a slight to Texans, but that is where I first heard of this practice as being socially acceptable (and that's what my Texan friends called it). In short, it means you don't flush after every use of the toilet, only after depositing something other than just urine. They even have a poetic memory aid for the practice:

> If it's yellow,
> Let it mellow.
> If it's brown,
> Flush it down.

A simple rhyme that could probably reduce your household water use considerably.

## Aerators / Flow Reducers

Install water-saver/aerator faucets on your taps. They really do reduce the amount of water you use. Some varieties provide additional functions, like a spray setting or the ability to angle the water flow (handy for rinsing the sink).

Use reduced flow showerheads, especially the kind with an in-line shut-off (also called a pause feature). These can cut water use by 50-70% (including hot water) over conventional shower heads, but the increased water pressure and aerated spray make it feel as though you are still getting about the same flow on you. The in-line shutoff permits you to stop the water flow completely when not required (e.g., while lathering), without affecting the temperature of the water (ratio of hot to cold in the mix) when you restart it. The shower massage type heads are very nice, but may lead to increased time in the shower because they feel so good. I'm a fan of the showerheads with short hoses built-in; very handy for cleaning up the shower enclosure.

## Turn It Off

Many of us let water run even when we don't need it. When washing our hands, turn the water off while soaping up, lathering and scrubbing, until it is time to rinse.

While in the shower, during the time we are soaping up, scrubbing or lathering up the shampoo, leaving the water running is actually counter-productive. So shut it off until you need it again.

When brushing your teeth, the water does not need to be running except to wet the brush, clean the brush, and rinse your mouth when you are finished brushing. A cup of water is usually more than sufficient for the whole procedure.

Sometimes though, you turn off the tap, but the water doesn't quite completely stop. It's the dreaded drip!

## Drips and Leaks

Fix them! Now!

Do you think they are going to miraculously fix themselves?

It's usually a matter of replacing a simple washer in a faucet, or a flapper in a toilet. Figure out what you're going to need and get it (parts and tools). Turn off the water at the main entrance to the building, or at a nearby cut-off (if you are so fortunate). Drain the system as necessary, then take the problem apart, fix it, and put it back together. Borrow a how-to-fix-it book from the library if necessary. If you're really intimidated, it's time to phone a friend.

If it really is major work, then hire someone qualified to do the job. Again, it's not going to get better with age. Leaks can lead to mould and rot if left unattended.

Until you do get the dripping faucet fixed, at least catch the water in a container. You can use it to water plants, wash vegetables or in the toilet tank when you flush—at least it's not simply going down the drain as a complete waste.

## Cut-Offs

Speaking of cut-offs, when the opportunity presents itself, install them. When (not if) plumbing problems arise, cut-offs are your best friend. Instead of having to shut off the water to the entire building, you can isolate the problem and keep the rest of your plumbing functional. Suppose it is 6:30 on Sunday evening, and you are putting the finishing touches on dinner, when a freak twist of the kitchen faucet erupts into a geyser over the kitchen counter. Would you rather be able to shut off the water flow to the kitchen sink and continue largely as normal, or be without any water or working toilets until sometime Monday? Cut-offs make the difference.

## Don't Water the Lawn

Grasses have grown throughout most of North America before the suburbs were surveyed and populated. They'll do so again without our hoses, sprinklers, and irrigation systems. You may wish to switch to a less thirsty type of grass or other ground cover (e.g., clover) to maintain the green colour.

Better still, get rid of your lawn altogether (and your lawnmower with it). Replace it with a garden, which can provide you with fresh food and flowers.

## Drip Irrigation (Not Sprinklers)

Of course, your garden may require watering for the plants to survive. However, they probably won't drink as much as a lush lawn. Drip irrigation systems water just the plants you want, and not the areas between, which helps control weeds as well.

## Wash Your Car Less Often

It's not like the car is going to grow because you water it.

When you do need to wash your car, do it with a bucket and sponge, not a hose. Use a mild detergent, which is better for your car's paint, your skin, and the environment.

If possible (and you still have some), wash your car on the lawn. Let the water you do use soak into the ground, instead of into the nearest sewer.

Having covered some ways to reduce your actual water consumption, let's turn to some options for alternative sources for water supply.

## Rainwater Harvesting

This may not be for everyone, for example, if you live in a desert, or don't have a roof. Otherwise, there is probably plenty of rainwater available to you to make harvesting worthwhile. If you do live in a desert, there are atmospheric water harvesting techniques that work. Research air wells, dew ponds, fog fences, and atmospheric vapour collection. Personally, I'm not a fan of the plastic sheet/solar still method of collecting water in a dry environment—it has not worked well for me. However, anywhere I have lived, there has been plenty of conventional precipitation on an annual basis.

Let's consider a typical suburban scenario. The property is nicely groomed, with a lawn, shrubs, some flowers, and possibly a small vegetable garden. And the homeowner is out there most days, tending to her yard, watering it, unless it has rained recently. Watering is pretty straight-forward. The garden hose is hooked up at the back of the house, brought around the side to the front, tripping over at least one downspout from the eavestroughing, to the nozzle. With this, the homeowner is laying down a daily barrage of water to keep the soil moist and the plants (including nearby weeds) healthy.

The water being used fell as rain a few days earlier, provided by Mother Nature. It was collected by a series of drains or creeks and ended up in the municipal water treatment facility. There it was treated with chemicals, and agitated, and pumped about, and filtered and settled, until it was pumped to a reservoir and eventually into the water supply system. Then it was pumped through the water supply labyrinth until it arrived at the house, and finally the hose, to be sprayed onto the lawn or garden. That's quite a trip to end up back where it had fallen just days before. A lot of energy was expended in treating that water to make it drinkable, and pumping it back out to the suburbs, just so it could moisten soil. That energy shows up in the homeowner's water bill.

Suppose the homeowner had a rainwater catchment system. Instead of flowing off the roof, into the eavestrough, down the driveway and into the sewer system, this rainwater was captured in rain barrels or some other suitable container. Now, instead of using chemically treated water for our garden (that we have to pay for), we can use natural rainwater to sustain our garden, for free.

Implementation is usually simple. Set up a container that can be filled from the top and drained from the bottom with a tap or spigot. Make allowances for overflow. Make sure all openings are sealed or covered with screening, so you do not create a breeding ground for mosquitoes. I prefer my containers to be set up at least a metre off the ground, so gravity can assist with the movement of the water. I also try to have one barrel on the ground (part of the overflow cascade) from which I can easily fill a watering can by dipping the can into the barrel. The size of the container should be appropriate to how much space you have, how much collection area, and how long you need to store water between rains. Smaller containers (e.g., plastic barrels) can often be ganged or cascaded together, allowing you to increase storage capacity over a period of time.

Rain barrel on stand for gravity feed to water garden, fed from
household downspout

In climates with hard freezes, you will likely have to drain your systems annually, and refill again in the spring after the thaw, to prevent damage to your containers.

Since we started rainwater collection at home, our use of city water for the
yard has dropped to zero. We have also hooked up drip irrigation hoses into the
garden from the rain barrels, so watering the garden now takes only a couple of
minutes of effort on my part. I turn on the spigot at the barrels to the hoses, then
a few hours later, remember to turn it off.

At our cottage, we got more ambitious. Originally, our cottage (seasonal residence) drew its water from the nearby lake. The lake water was not used for
drinking or cooking, only for showers, washing dishes, and cleaning. We either
brought our drinking water with us, or replenished from a nearby community
spring. Over the years, I got tired of putting the foot valve in frigid water each

spring, and pulling it out again in the fall. If we put the valve out too far, the motorboats would cut the suction hose. If we put it in too close to shore, we got terrible water by mid-summer with the seasonal algae blooms. The hoses were a pain to maintain, and expensive to replace.

Eventually, I resolved to sever our connection to lake water. It turned out a well was not an option. Our lot is too small to meet the municipal regulations for a well (passed long after the cottage was built), plus our neighbours who had made considerable investments to drill wells were unhappy with the quality of the water they got from them. The more I thought about it, the more rainwater collection appealed to me. The final system is not in place yet (it will be based on a permanent cistern buried under the cottage to avoid freezing). However, the interim system, based on ganged barrels, is working to our satisfaction. The rain falls onto the cottage roof, into the eavestrough (salvaged from replaced eaves-trough projects), and then into a rain barrel cascade, which filters out the major debris via a fine mesh screen. From here the water is drained into the main storage. The main storage includes a foot valve, a four-foot section of suction hose (instead of almost four hundred into the lake) to the conventional pump and pressure tank.

Compared to the original deal, this arrangement is a joy! I certainly look forward to the completion of the final set-up, which won't even require the spring set-up and autumn draining of the interim set-up. However, for now, we get lovely, incredibly soft, clean-smelling and clear water in our sinks, the toilet and the shower.

Warning: once you have shampooed in rainwater, you won't want to go back to city or well water.

A final thought on urban rainwater collection and flooding. Consider the journey of a raindrop that falls on a conventional city roof without a collection system, versus the journey of a raindrop that falls into a forest or marsh, much like our property would have been before it was developed into housing.

First, our country raindrop. It falls onto a tree leaf, cleaning some dust from the leaf on impact. Then it puddles and flows down the leaf stem, then the branches, and finally the trunk of the tree, where it soaks into the ground at the base of the tree. Assuming it is not absorbed into the root system of the tree or another plant, it slowly migrates to a marshy area, where it pools for a while: hours, days, or weeks. With time, it may flow towards a creek or stream that drains the marsh, and from there wind its way to a major river over a period of hours to weeks. Assuming this specific raindrop is not absorbed by vegetation, it will likely take weeks to reach the major river.

Now, let's consider our city raindrop. It falls onto a roof, and slides down into the eavestrough. From there it slides to the downspout, and then is directed onto the driveway, and down to the street, where it streams its way to the storm water collection system. Possibly it missed a roof, and hit a yard landscaped for efficient drainage, or a driveway or street. It makes little difference. Save for a tiny portion that will be absorbed into lawns and gardens when the rain is actually falling, the whole arrangement is constructed to speed the rain into the storm collection system. From there it goes to the wastewater treatment system, at least until it overflows into a major river within or adjacent to the municipality. The whole trip likely takes mere minutes for the city raindrop, instead of weeks for the country raindrop.

In the past few years, our municipality has had increasing problems with flooding of basements and streets as the storm water system has been overwhelmed during heavy rainstorms, flowing back into household drains and street sewer systems. Why? Not because the storms are more intense, lasting longer or more frequent. It is because the area we are draining so efficiently is expanding as we continue to build more housing. The newer, smaller storm water collection pipes don't have the same storage capacity as the old tunnels that workers could walk through, so the new storm surges are delivered more quickly to the central treatment plants than before. Many millions of dollars have been spent compensating the flooding victims, constructing new storm water holding lagoons, and upgrading the storm water handling systems. From my reading, I have no reason to believe this municipality is unique in facing or responding to this issue.

Suppose we tried a different approach. Rain barrels—lots of them. Half a dozen for each household. Let's assume this scenario. Suppose the gully-washer of the decade is about to hit, after a dry spell. The ground is hard and dry. When the rain hits, it isn't going to soak in, it's going to bounce off and head straight to the drains and overwhelm the storm water system within a couple of hours.

If 100,000 households each have six rain barrels (1200 litres or 340 gallons) to fill. There will be 120,000,000 litres (34,000,000 gallons) of rainwater that won't hit the municipal storm water collection system during the storm period. By the time the rain barrels start to overflow, the ground will be wet enough to start absorbing some of the continuing rainfall. The collection system will be spared a huge amount of the potential volume, and what does come will arrive much more slowly. No urban flooding. No need for central storm water lagoons, because the same volume of water will be stored at the households. No need to treat or divert that volume of water. The homeowners have a source of free rainwater for watering their lawns and gardens, filling splash pools for kids, or washing cars. The volume of water stored in the suburbs will have a moderating influence on temperatures. This is undoubtedly far too radical for municipal officials to contemplate seriously, but you can do the set-up for your property (and reap the benefits) without any need for a citywide project.

In addition to the rain barrels, consider living driveways. Instead of asphalt, concrete, or paving stones, we could encourage the use of living driveways. The simplest form I have seen is to use pavers on their sides in a box pattern, leaving square openings for soil. Grass (or whatever you prefer) can grow between the pavers, providing an area where water can drain down instead of simply run off.

## Grey Water

Grey water is essentially any water used in your household that goes down a sink drain. If it goes down the toilet, it's called black water. Grey water systems are still in their infancy as part of new home construction. Grey water can be used to flush toilets or water household plants, lawns, or gardens. Most municipalities do not have standards for them, and to date only the eco-pioneers have had the foresight and desire to implement them. So, you probably won't come across them in the suburbs for some time to come. From my reading, it appears that a retrofit is a significant undertaking, so I won't delve into the subject of retrofitting grey water systems here—the payback isn't there to justify it.

However, that does not prevent us from using grey water to our benefit, even without the installation of the formal plumbing works. There are varying shades of grey water. Water drained from cooking is usually fine for watering plants (inside or out). Water used for washing our hands is usually just fine too. In some energy efficient homes, the bathroom sink drain is plumbed into a reservoir that can be used to fill the toilet. If you are keeping the bath water in the tub to extract the heat from it, once it has cooled, it is also fine for watering plants. A little soap is not going to hurt your plants.

Using the last rinse water from one load of laundry to be the first water in the next load is a form of grey water re-use.

With a little imagination and some common sense, you will likely find other grey water uses. The biggest drawback is usually the need to store and transport the water.

# 27

# *Personal Energy Production*

Let's say you have worked your way down to the lowest energy consumption at which you can function without sacrificing your quality of life, and are ready to address how to meet your remaining energy demands in a sustainable way. Just as there are many things you could do to reduce your energy demands, there are many ways you can produce the energy you need.

## *27.1. Low Tech Solar*

If you have not guessed by now, this is my favourite. True, the sun doesn't shine at night, and there are days it doesn't put in much of an appearance around here, either. When it is available, however, it's absolutely free, and collecting it is generally cheap and easy. Here are the main ways I use low tech solar energy. I'm sure there are many others.

### Solar Space Heating

It really can be as simple as drawing back the curtains and letting the sun shine in. That's passive solar heating in a nutshell. From there, you can get increasingly ambitious as time, space, funds, and motivation permit.

You can build a solar closet or simple solar (Trombe) wall that acts as a solar heat collector, and either bring it directly into the building structure or into some kind of thermal mass for providing heat later. Personally, I'm not a big fan of the Trombe wall. If you are going to do this much work, why not make the small extra effort to create an active solar heating system that will drive the heat into the structure, instead of losing much of it back outside at night through the glazing?

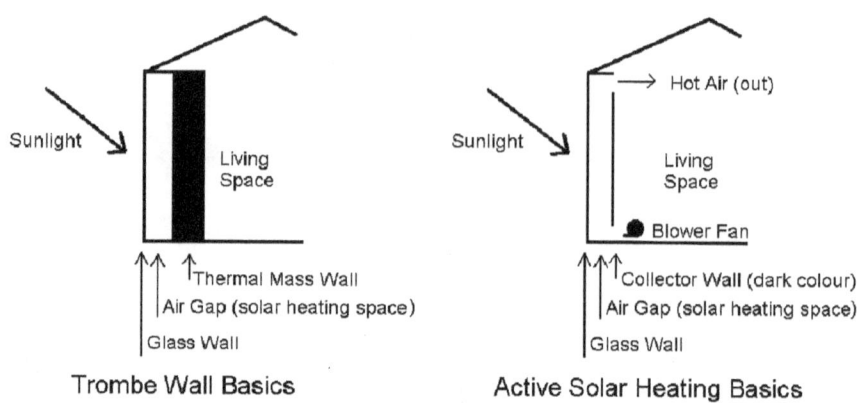

Figure 3

In the illustration (above left), you can see a Trombe wall collecting the heat energy into the thermal mass (wall). At night, heat radiates back outside (where it provides you with no benefit) at least as easily as into the living space. With an active solar heating system (above right), the heat is pushed into the living space, where it is easier to retain when the temperature cools off outside. The drawings above are very basic, and the active solar heating systems include refinements that increase their effectiveness.

You can build a sunspace that will act as a collector. This is a space that is not intended as a living space, and should not be heated. The objective is to collect energy from this space, not use energy on it. Ours is a seasonal structure (greenhouse) that is connected to our house via the patio door to the back yard. When the temperature in the sunspace is higher than it is in the house during the heating season, we open the patio door to let the warmer air flood in. We keep the thermostat set at 19°C (66°F) during the day when we're at home. That's our base comfort temperature. That may sound cold to many people, however, we wear slippers and sweaters, and our house is no longer drafty. On a nice, sunny, winter day, the sunspace can reach temperatures of over 50°C (120°F) without difficulty. When we open the patio doors, the sunspace can drive the temperature on the main floor to over 26°C (79°F), and the sweaters and slippers come off.

Active solar heating consists of adding some components that push the heated air or fluids around. In our house, we use three solar heating panels that I bought on ebay. They cover almost 10 square metres (about 100 square feet) of our sun-facing exterior wall. There is 6-inch ductwork to carry cool air from our basement floor to a blower motor, through the panels, and back to the basement. The brain of this system is a differential thermostat that determines whether the panels are significantly warmer than the basement. If they are, the system turns on the blower motor and opens the damper. When the temperature in the panels drops below the temperature indoors, the controller shuts things off. The blower draws a bit less than 100 watts, and I estimate the panels deliver about 5,000 watts of heat on a clear, sunny winter day. We shut the system off in the spring, and plug it back in again in the fall. Other than that, it's automatic. I have enjoyed many hours basking in the heat of this solar collection system while writing this book, as the outlet is over my desk.

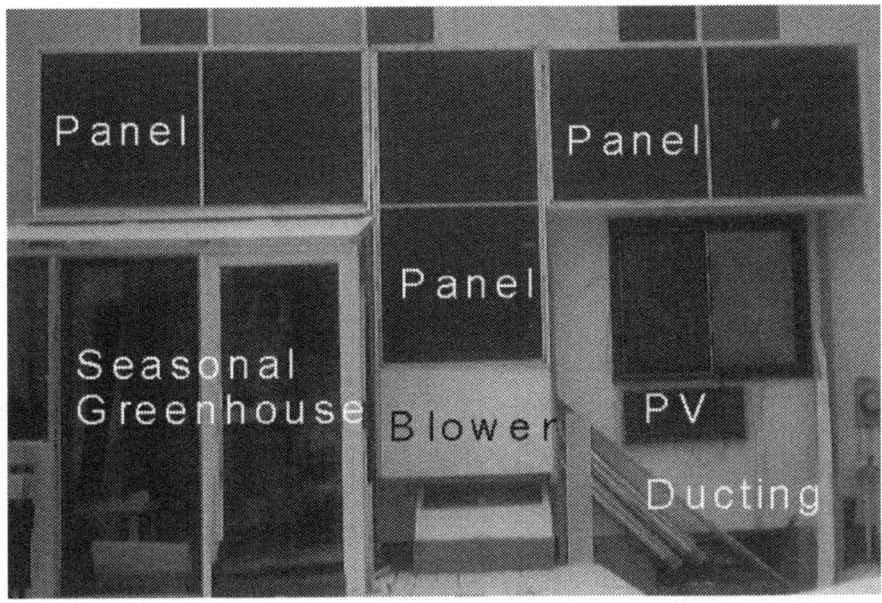

Solar heating panels, PV panel, and greenhouse on author's house

When old Sol is really on the job, there have been days our furnace did not run at all even in the dead of winter, and outdoor temperatures may range from -30°C (-22°F) at night to -15°C (5°F) during the day. Ours isn't a custom-designed, high tech solar house. It is a conventional row-house end unit built before the 1973 energy crisis with no thought to energy efficiency. We have simply upgraded it over time with small investments in better weather sealing, more insulation, and some simple and affordable solar collection gadgets. We haven't even finished upgrading all the windows yet.

Here are some additional tips to maximize solar heat gain:

- Make use of passive solar energy by opening window coverings on sun-facing windows when the sun is shining. During the winter, in a typical North American residence, this can contribute about 3,000 watts of heat energy while the sun is shining.

- Have the sunlight shine onto dark, matte surfaces, and increasing thermal mass enhances the heat energy production and retention.

- Clean your sun-facing windows in the autumn to allow as much energy as possible to flow in during the heating season.

- In new constructions, make sure larger windows are sun-facing, especially towards morning and mid-day sun. Walls facing away from the sun should have smaller or no windows.

For active solar heating systems, do your research (consult the Internet and the many books on the subject), or consult with an energy expert or architect knowledgeable on the subject.

## Solar Water Heating

Most of us heat our water by burning oil or natural gas, or by converting electricity into heat. Heat is the lowest form of energy. While it may make sense to use one of these high quality energy sources for the finishing touches on heating our water (to ensure it is maintained at the desired final temperature), there are lower quality forms of energy that we can use for the bulk of our heating requirements. My personal favourite is solar energy.

Solar heating of water is hardly something new. In many countries it is common (e.g., Greece, Israel). The technologies available range from simple—the sailor's black water bag—to complex—dual loop systems with computerized controls.

For simple systems, create a flat black vessel, fill it with water, and put it out in the sun to collect infrared energy. Wait a little while, and you will have hot water. The vessel can be as simple as a black plastic bag or a black hose or pipe lying in the sun, or it could be a sophisticated system of collector panels and reservoirs. Again, your choices are only limited by your funds and ambition.

There are few, if any, solar water heating systems that can provide all the hot water you will want year-round in most of North America. The U.S. sun-belt and Mexico are possible exceptions. For most of us, we should think of solar water heating as a pre-heater to another heat source that will supply the additional required energy to provide hot water. I know there are folks who are prepared to live with cold showers, but my concept of quality of life includes hot running water, every day.

The system I installed is a simple, seasonal, batch solar water pre-heater. I built this from the inner tank of a 60-gallon hot water heater that was declared surplus from a water heater rental company. I enclosed it in a box made of scrap plywood. I spray-painted the tank flat black. The box was insulated inside on all sides but the sun-facing side, the tank inserted, secured and plumbed, then covered with a surplus patio door (double pane), and sealed with caulking. We used a reflective covering over the insulation in hopes of getting a bit more sunlight onto the tank. A cheap, remote-reading thermometer was installed on the side of the box, with the sensor inside, out of the sun and midway down the tank. If I were to do it over again, I would put the sensor at the top of the tank, near the hot water outlet, so that I would know the temperature of the water actually leaving the solar heater, not the average temperature. The tank was plumbed into the cold water line that fed the existing, conventional hot water tank using screw-on garden-hose type fittings. Simple and cheap. No dual loops, heat exchangers, propylene glycol, or extra reservoirs.

Our unit includes a conventional over-pressure, over-temperature relief valve. It is undoubtedly unnecessary in our climate, but it was available and an additional safety feature. If you include this item in your solar water pre-heater, I recommend storing it (when not in use) so that the glass surface is not facing the sun, or covering the glass with something opaque to prevent the sunlight from heating the box. Without the moderating influence of the water's thermal mass, the temperature inside the box can exceed the rating on the relief valve, requiring it to be replaced.

Our batch solar water heater
(Large black tank in insulated box covered with salvaged patio door)

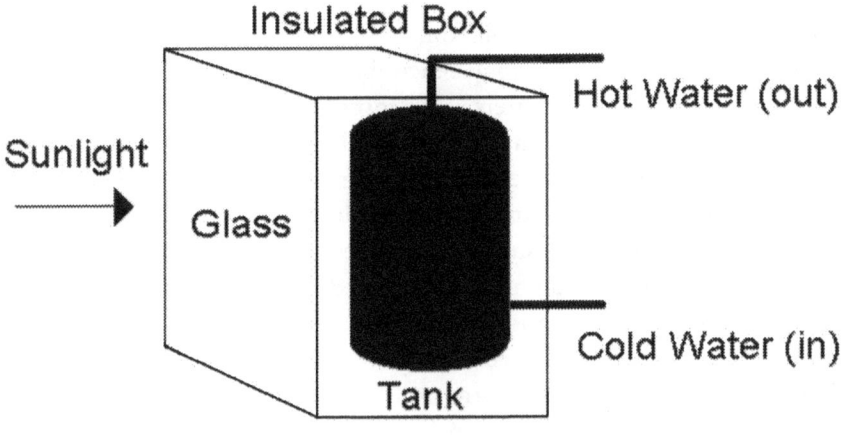

Figure 4
Solar water pre-heater concept diagram

In the spring we put the solar water pre-heater in line (aiming the glass at the sun around noon), and in the late fall we take it out of line and drain it (to avoid freezing damage). So long as the water in the solar tank is above 10°C (the temperature we get from the municipal water supply varies from 6° to 10°C depending on the time of year), we leave the solar water collector in operation.

Does it work? Yes. Our water heating bill (natural gas consumption) dropped from about $20 per month (about 82 cubic metres) without the solar pre-heater to effectively zero from mid-May to the end of October with the pre-heater. According to our bills, we used only $0.75 worth of natural gas during the first full season (2003) we had the solar pre-heater installed (3 cubic metres). That's about ½ cubic metre per month, instead of 82. This is a total savings of about $120 over six-months. Based on the cost of the materials used, the payback was less than a year. Assuming this device lasts just 10 years, the return on investment is at least 1000%. Natural gas prices have already risen by over 50%, so the return is even better than stated. It takes about an hour to set the tank up in the spring, and another 90 minutes in the fall to disconnect, drain, and protect it.

After considering various options, we felt this one made the most economic sense for us. It reduces our fossil fuel consumption, our environmental footprint, and our energy bill, even though we are limited to using it only about half the year. If you live in a warmer climate, presumably you could derive more benefit than we do from this simple approach. You may prefer a more sophisticated system that delivers benefits for a greater part of the year, especially if you qualify for financial incentives to do the installation.

When it comes to implementing renewable energy on the home front, efficiency doesn't matter. What matters is the cost of delivering a unit of energy over the life of the device. You can buy a commercial dual-loop solar water heater that meets 95% of your hot water requirements for $4,000 or more, which we will assume lasts thirty years. You can also build a simple solar batch heater for about $120 that will meet 50% of your hot water needs, which we assume lasts just 10 years. The latter system will do better than stated in warmer climates than ours.

The more expensive system is more efficient. If it displaces natural gas as the heating fuel, and it costs $240 a year for natural gas water heating before implementing a solar pre-heater, the return on investment for the more expensive system is about 80%. Not a great return financially over a thirty-year period. The return on investment on the lower priced alternative is about 1000% (probably high as the cost included nothing for labour). That is almost 100% per year over the ten-year period. The more expensive option needs higher priced conventional fuel to become financially attractive (not a bad bet, but not quite a sure thing yet). Just remember, efficiency is not as important as meeting your requirements and the return on investment. This is an important concept to remember when we evaluate projects.

Most people need to start thinking about sustainable energy in a small, low-risk way. I think the batch solar water heater is a classic example. What do you have to lose? Your water heating bill? If you build it and then decide to install a more capable sustainable water heater in the future, you can always give away (or sell) your batch water heater to someone else.

Beyond typical hot water demands (showers, baths, washing dishes, and so on), another significant hot water use for some people is heating swimming pools and spas. Here again, solar water heating can do a great job. The existing circulation pumps are usually sufficient to move the water through collectors to provide the desired heating. The major additions are typically the collector panels and supports, the additional plumbing run, a differential thermostat controller, and the valve(s) it will control. The additional electrical energy cost is trivial because the pump is running anyway. The heat from the solar collector can extend the useful pool season by at least a month at each end.

There are plans for simple solar water heaters on the Internet and in several books on solar energy. There are also specialists that install the more complex systems. You can likely find a supplier in your local phone book or other advertising media.

## Solar Cooking

We spend a significant amount of energy on cooking food. Typically, we in North America use electricity, natural gas, and propane (and to a lesser extent wood) as our primary cooking fuels. We also use barbecues and open fires for some cooking in some settings.

Some visionaries have seen a more sustainable way: cooking with the energy of the sun. While it is not viable as a sole means of cooking in most areas, many of us have multiple means of cooking at our disposal (ranges, ovens, microwave ovens, toaster ovens, crockpots, barbecues, etc.) For many of us, adding one more is more likely to be entertaining than burdensome.

There are a variety of solar cookers, each with their own characteristics, strengths, and weaknesses. Before investing in a commercial solar cooker, learn what kind of sun you usually have (bright, strong and continuous, or hazy or frequently interrupted). Then find which type of cooker works best in those conditions (box cooker, parabolic, funnel, bag, dome, with or without reflectors, etc.). The simplest are quite inexpensive, including the Heaven's Flame design (a solar oven),[1] and those based on simple reflectors and clear plastic cooking bags. These typically act like a cross between a small oven and a slow cooker (crock pot). There are several Web sites with good information on solar cooking providing additional information to get you started. Solar Cooking International hosts solarcooking.org.[2] An Internet search will find many more, such as solarovens.org.

If I may promote one specific product, it is the Solar Sizzler. Most people that experience solar cooking eventually want a cooker that has more power and cooks faster. The traditional solution is the parabolic reflector. However, these are difficult for most of us to build well, and awkward to store. The Solar Sizzler folks have solved that for us, by making a clip-together, highly reflective, parabolic reflector for a very reasonable price. It breaks down into a small, lightweight package, suitable even for backpacking. You can find them on the web at: http://www.solarsizzler.com.

Whichever solar cooking device you choose to use, try to have fun with it. I certainly have. And have a backup plan to complete the cooking, in case the storm clouds roll in at an inopportune time.

---

1. *Heaven's Flame Solar Cooker* by Joseph Radabaugh
   http://www.backwoodshome.com/articles/radabaugh30.html
2. Solar Cooking International Web site
   http://www.solarcooking.org

Warning: solar baked bread can be habit-forming, and bad for your waistline.

## Greenhouse

While the greenhouse effect may be bad for us at the planetary level, it can be quite beneficial at the residential level. While not all of us can afford the space or funds for a full-size, permanent greenhouse, that doesn't mean we can't benefit from smaller versions.

One small, simple version of the greenhouse is the cloche. A cloche is a glass (or sometimes plastic) dome that can be placed over a single plant. This provides solar heating to the area under the dome while the sun is shining, and provides protection from frost when the temperature drops. Cloches may have to be removed daily to prevent cooking the plant.

The next level up is the cold-frame. This is essentially a window that covers a box. The plants are placed inside the box, and solar heating is again used to benefit growth, while the enclosure protects the plants. Most cold-frames include the ability to open the glass lid to prevent over-heating inside the frame. This can be automated, but is usually done manually. Cold-frames can be integrated into raised or box gardens. It's a great way to re-use old windows.

Our greenhouse is a bit of an over-sized cold-frame. It is a temporary structure we put up on our rear deck in the fall, and take down in the spring. It serves several functions. For most of the winter, it provides an additional level of insulation over our patio doors. On sunny days, it acts as a solar closet, and we let the solar heated air enter the house through the opened patio doors. In late winter to early spring, we use it to start our plants for the garden that need the early start (e.g., tomatoes). We have also experimented with trying to grow in the greenhouse through the winter, but as we do not heat it and our winters are harsh, nothing we have tried (including hardy winter varieties of greens) have survived. However, experiments are likely to continue in future winters.

A full-scale, permanent greenhouse takes some planning to do right, and needs provisions for heating other than from sunlight in order to protect crops from freezing temperatures (assuming a climate where a greenhouse would actually be of benefit). Greenhouses with removable insulation (such as pumped bubble-foam between double walls) are an alternative to conventional heating for maintaining greenhouse temperatures.

## 27.2. Biofuels

In addition to the use of biofuels in transportation (biodiesel, vegetable oil, ethanol, methane, wood gas, etc,), biofuels can also have a role in stationary applications, like heating fuels.

Biofuels covers a range of possibilities. The common link is that we can grow more, which makes these energy sources sustainable. With water and care of the soil, biofuels become energy stores for solar energy (via photosynthesis). We will still need to use the energy carefully, so that we don't end up denuding our planet of forests or arable land. Conservation (i.e., insulation, weather sealing, reducing distance traveled, etc.) still comes first.

A common criticism of biofuels that has long intrigued me is that we can't grow enough biofuels to replace our current use of fossil fuels. Why would we want to? Surely, if the impetus for looking at biofuels on a mass scale is to replace increasingly expensive fossil fuels, then we should be looking at ways to reduce our energy consumption, as well as displace the dependence on increasingly scarce fossil fuels. If the scarcity becomes acute, well, we'll pretty much take what we can get, won't we?

If fossil fuel scarcity is not the driving argument, then what does it matter if biofuels can replace all fossil fuels, so long as those of us who choose to use biofuels can be fairly sure of a steady supply sufficient for our own needs?

Another criticism of biofuels is that there are a variety of non-interchangeable biofuels, and the consumer has to commit to one or another (e.g., wood vs. biodiesel). This strikes me as an odd point in a world where we routinely choose between wood, electricity, oil, or natural gas for heating; between gasoline or diesel engines/fuels for transportation. Our utilities choose between hydro, coal, natural gas, nuclear, and other primary sources for the production of electricity. Why is this perceived as a weakness for biofuels, when it doesn't seem to be an issue for the current conventional energy mix?

The real beauty of all biofuels is that we can make our own, at a household scale, with relatively simple technologies. And when you make your own energy, you really come to appreciate it much more.

### Wood

The traditional favourite, wood has been with us for millennia, and remains a popular fuel throughout much of the world. Its primary energy function is to produce heat for heating homes, cooking, and heating water. As wood has other valuable uses (e.g., furniture, construction material, paper-making), in some situations wood scraps are readily available. A fairly recent variant on wood scraps is pellet stoves. The pellet stove is a cross between the conventional (air-tight) wood stove, and a fueled furnace.

Woodstoves are taking a bit of a beating in the mainstream press of late. So let's be clear on a couple of key points. First, burning wood creates no more carbon dioxide than is released when the same amount of wood rots on the forest floor. Burning simply releases the carbon dioxide faster. If trees grow to replace the wood being burned, the overall impact should be carbon neutral, over time. Second, modern woodstoves are not a major source of pollution in North America. I have seen the smoke plume from major forest fires, and that's definitely a problem. However, modern woodstoves that are set up to maximize heat from the wood consumed produce very little smoke or ash up the chimney. I've watched. It is not woodstoves that are producing pollutants like sulphur dioxide or mercury. For that you need to look past the smokescreen at the more likely culprits—coal-fired generating plants.

The old adage about wood heating twice is true. Once when you cut (move, split, stack) it, and again when you burn it. If you are planning to buy a woodstove, do yourself a favour and invest in efficiency and quality. An efficient woodstove will save you a lot of wood, and the associated effort or money. A poor woodstove is one you won't use, which makes it a poor investment.

## Ethanol

People have been making ethanol on a small scale for a long time, though it may be better known as moonshine from stills. In North America, the toughest part about brewing ethanol for your own use as a fuel is getting the necessary permits from the various levels of government. Seems those suspicious folks paid from your taxes are afraid you're going to drink the stuff, and might avoid paying taxes on the alcohol you would otherwise buy (and pay taxes on). That's just crazy! The stuff's too valuable to drink! Instead, we can use it to power our spark-ignition engines, as a cooking fuel, or for heating (in alcohol stoves).

The accepted practice is to mash fruit or other high sugar-content vegetable matter (beets, corn, sugar-cane), ferment the mash, heat the storage vessel to evaporate the alcohol, and then condense the vapour to liquefy the desired end product for more convenient storage. I understand some government jurisdictions will supply you with designs for stills when they grant you a permit for producing ethanol. They may also require you to denature it (contaminate it with a toxic substance, e.g., gasoline), to ensure you don't sneak a sip when they're not looking. Nasty stuff, that gasoline. Pity you have to keep some about just for this purpose. Other denaturants could include methanol.

There are plenty of good books on producing alcohol/ethanol, as well as sources on the Internet. Therefore, I'm not going to use up space here describing it in detail.

I expect that cellulosic ethanol production will remain beyond the means of the individual producer for some time to come. The debate continues as to whether the backyard distiller can produce dry ethanol that embodies more energy than was used to produce it.

## Biodiesel

Biodiesel is essentially vegetable oil (occasionally animal fats) that has been put through a chemical process (transesterification), to create a fuel that will run in compression-ignition (diesel) engines. Typically, the chemicals required are methanol and lye. There are multiple recipes, and some use different chemicals. The whole process is not much different in principle or complexity from old-fashioned soap making.

Biodiesel can be used as a replacement for fossil heating oil (blended or as a direct substitute), as well as a substitute for diesel fuel.

I have made biodiesel from used cooking oil (UCO) in small quantities, and have been very satisfied with the results. Biodiesel is typically very pale amber in colour (depending on the original oil) and smells like cooking oil.

We recently acquired our first diesel vehicle. To date we have run B20 in it regularly. Between the lubricating quality of the biodiesel and the synthetic oil we are using, the unique diesel top-end racket is almost eliminated, and some folks don't believe it's a diesel when they hear it. Before long, I hope to be ramping up production of biodiesel for our own use. Not only is it a good way to recycle a waste product (used cooking oil), but it also constitutes a zero-GHG combustible fuel. This is because the plants that will be grown to produce the next crop of vegetable oil will extract more carbon dioxide from the atmosphere than is produced from the combustion of the re-processed vegetable oil.

It's an art to make quality biodiesel from vegetable oil and requires practice. You don't want to throw garbage into your fuel tank. Check your vehicle warranty for details before proceeding; many permit a small proportion of biodiesel blended with regular diesel. Then, do your homework. Brewing biodiesel is not for everyone. It takes space, time, patience, and attention to detail.

Certain cooking oils have different characteristics, which are reflected in the biodiesel produced from them. For example, the gel point for canola oil is lower than the gel point for most other vegetable oils and animal fats, which makes it more valued for use in areas that experience colder temperatures.

There are several books and many Web sites on the subject of biodiesel. I have spent time on many of the Web sites, and read more than one of the books. In my opinion, the single, best resource is the biodiesel library at the Journey to Forever Web site:

http://www.journeytoforever.org/biodiesel.html

As Keith and Midori (Web site hosts) would say, start here:

http://www.journeytoforever.org/biodiesel_make.html#start

Warning: The Journey to Forever Web site is quite large, and covers much, much more than biodiesel. This Web site can be addictive.

Sample jar of biodiesel made from used cooking oil by the author

## Vegetable Oil

A variant on biodiesel is the use of straight vegetable oil in diesel engines. In general, this approach requires modifications to the vehicle fuel system, notably a second fuel tank and a switch between the tanks. Diesel engines do not start well on vegetable oil. Therefore, the standard practice is to start the engine on regular diesel fuel (from the smaller tank), and once the engine has come up to temperature, switch over to the straight vegetable oil. The really tricky part is remembering to switch back to regular diesel for about five minutes before shutting off the engine, to ensure that all the regular vegetable oil has cleared the engine before shutdown.

This approach is not particularly beneficial or practical if most of your trips are short (fifteen minutes or less). The oil becomes increasingly viscous as temperature falls, making it less attractive in cold climates. Some kits address this with fuel line and tank heaters. Personally, I don't understand the attraction, but then I don't do a lot of long distance driving. The technology has a goodly number of adherents.

Of course, if you have the land (or can rent it), you can grow your own oil crops (e.g., canola), harvest it, and process it into oil for your own use. You could also process the vegetable oil to make biodiesel.

## Methane

Certainly, if you have the means to produce your own methane (or acquire it affordably from a local source, e.g., a landfill) and a means of using it, say a modified natural gas heating system, hot water heater, or compressed natural gas (CNG) vehicle, then this would seem an appropriate means of using a sustainable energy source. You will need an appropriate pump to fill a pressure tank such as those found on CNG vehicles. It has the added benefit that burning the methane will produce carbon dioxide. While carbon dioxide is a greenhouse gas, it is much less potent than methane when it comes to greenhouse gas emissions.

This may be particularly feasible on farms with appreciable amounts of animal manure. The manure can be placed in a biodigester to extract the methane for use as a heating or engine fuel, and the remaining material subsequently composted for crop nutrients.

Methane can be stored in inflatable bladders or balloons or even old inner tubes, if more is being produced than is required on a seasonal basis. If the appropriate equipment is available, it can also be stored in pressurized tanks, just like CNG. Natural gas is mostly methane.

Vehicles running on CNG can presumably be run without modification on compressed methane produced from a biodigester, landfill gas, or other sustainable source.

## Corn

A variant on the wood pellet stove is the corn stove. The dried kernels are fed into the stove and burned to produce heat. It's easier to grow your own corn than to make your own wood pellets. Compost the rest of the corn plants. Speaking of compost…

## Compost

I'm sure you're thinking, compost is not a fuel. However, it is a potential heat energy source. As materials compost, the bacterial action produces heat. The temperature in the middle of a compost pile can easily exceed 50°C (120°F). Some individuals have created compost piles with plumbing run through them to extract some of that heat for heating water or buildings. In times past, it was common to put the compost pile against the outside wall of the house to benefit from some of the heat being produced.

I can remember when a great aunt of mine used to bank her house with seaweed in the autumn—she put piles of the stuff all around the foundation of the building. Originally I thought this was a pretty lame form of seasonal insulation. However, I later learned the decomposing seaweed over the late fall and winter helped keep the foundation warm, thus helping to warm the building. In spring, the remains were shoveled on to the garden or the compost pile.

So what can I tell you? Compost! Not only can you produce heat, it's great for your lawn or garden, reduces waste haulage and landfill utilization, and it reduces greenhouse gas emissions relative to landfills. Aerobic composting leads to the production of carbon dioxide, a GHG. However, anaerobic landfilling of plant matter leads to the production of methane, a much more potent GHG.

## 27.3. CHP

Combined heat and power (CHP) is also known as co-generation.

Doesn't it seem absurd that in many parts of North America we use very large, centralized heat engines (nuclear reactors, coal-fired, or natural gas-fired generating plants) to produce electricity at maximum efficiencies ranging from 30% to 60%, in order to run a significant number of appliances that are used for cooling (e.g., air conditioners, refrigerators, freezers)? The generating stations use a significant amount of energy just to dispose of the waste heat, thus warming the local environment. They're using energy to throw the heat energy away.

The cooling appliances are probably doing an appreciable amount of their work just to offset the waste heat that is being produced in massive quantities by those large centralized heat engines. All of this is happening in the context of a planet where global warming is considered a significant climatological and environmental issue.

Even in those locales where the heat energy might be valued, it is seldom harnessed for effective use (e.g., pre-heating of water, local space heating, or district heating). This is because, to a large degree, we not only use the wrong devices to produce electricity (heat engines), but even where those are appropriate, we put them in the wrong places.

It would make more sense to put small heat engines in our homes in place of furnaces as CHP devices. Instead of using a conventional furnace at 70% to 90% efficiency to produce heat, we could install small heat engines that would produce almost as much heat per unit of fuel consumed, and also produce electricity via a coupled generator (or alternator). You may prefer to house such a unit in a shed, rather than the basement, due to noise and exhaust issues, and run heating lines and wiring to the residence building. Such small engines could be run on the existing fuel infrastructure. Diesel engines (compression ignition) can be run on heating oil. Spark ignition engines can be run on natural gas. The heat can be captured from the engines to heat domestic hot water and provide space heating. The electricity can be used to operate regular appliances or charge batteries to supply electricity as required later, or be fed into the grid.

CHP facilities report combined overall efficiencies in the same range as conventional furnaces, while producing at least some of the energy output as a higher-quality energy (electricity), instead of just low quality heat energy.

As our energy intelligence matures, we can switch these engines over from the existing fossil fuel infrastructure to biofuels (either centrally or independently). The diesel engines can run on biodiesel or even straight vegetable oil (though two-tank operation—diesel and vegetable oil—is recommended). The spark ignition engines can run on methane gas, a direct substitution for natural gas, or ethanol, already used as a gasoline substitute.

Much more affordable than a household fuel cell, with similar efficiencies.

## 27.4. Wind Power

Wind power is usually divided into two broad groupings: commercial wind power development (big wind) and units aimed at consumer size requirements (small wind).

Big wind is characterized by wind farms with tens or more of turbines in proximity, sited at locations with superb wind energy characteristics (speed, consistency). Each turbine can generate a megawatt of electricity or more. Companies like Vestas and General Electric are major players.

Small wind is characterized by much smaller wind generators; they are typically installed in ones and twos, and operated by homeowners to provide energy primarily for their own use. In some cases, net metering agreements with local utilities may permit them to put surplus production on the grid, and possibly even be paid for it at either wholesale or retail rates (net metering). Once again, I won't cover this in any depth, as there are many books available that cover the subject well.

However, even if a wind turbine is impractical for you in your current circumstances, that does not mean that you cannot benefit from the use of wind power and support its use. There are wind power projects that are financed by means of support from private individuals via Green Tags.

Green Tags provide the purchaser with the assurance the electricity that they consume from their utility, no matter how dirty the original source, will be offset by the production of electricity from a sustainable source being fed into the grid. Personally, I support wind power projects like the Port Albert wind power project at Lake Huron via Green Tags Ontario[3] and via the Pembina Institute. There are other green electricity suppliers throughout North America.

More information on wind power is available from organizations like CanWEA[4] and AWEA[5].

## 27.5. Micro-Hydro

The big catch to micro-hydro is having access to a substantial and reliable flow of water. Few of us have a reliable stream running through our yards. Therefore, I am going to consciously gloss over this possible home-scale energy source in this book. However, that doesn't mean you should not pursue low head hydro if you have the resource. There are several good books on the subject. It is not simple to do micro-hydro well and extract the maximum benefit from it. There are multiple technologies, each with their own strengths and weaknesses. Study before committing, and seek expert help so you can get it right the first time.

---

3.    Green Tags Ontario
      http://www.greentagsontario.com
4.    Canadian Wind Energy Association (CanWEA)
      http://www.canwea.ca
5.    American Wind Energy Association (AWEA)
      http://www.awea.org

# 27.6. Solar (Photovoltaic) Panels

Photovoltaics produce direct current (DC) electricity from light, usually sunlight. They are the photo-op (pardon the pun) alternate energy source of choice for those looking to make a statement when cost is not an object. Photovoltaic panels are still too expensive to compete with commercial power for those of us already connected to the grid. The key to making most alternate energy sources (including photovoltaics) effective in any application is to focus on reducing the amount of power you will need to produce. I am not against the use of photovoltaics (having some myself), however, in my opinion, they should not be the first power source to be considered for residential or commercial power supply. There are often better choices available, economically and environmentally.

Photovoltaic panel on rear deck of author's electric car

When photovoltaics are truly cost-effective relative to conventional fuels, I expect the electrical utilities will buy them in bulk and sell you the power themselves. After all, that's their business. However, undersized grids and incentives targeted at consumers may make PV attractive to you before your local utility.

There are several factors to consider before plunging into photovoltaics. First, do you have good solar potential in your location? Is your proposed installation sun-facing? Does it have unimpeded access to sunlight during all daylight hours? Does the site get a lot of sunlight, or is cloud cover, fog, haze, or smog a regular feature of the local climate (as is the case for most urban Ontario locations)? Will you continue to receive an effective amount of sunlight as global climate change continues to create more sun-blocking gases and clouds in the atmosphere? (Really! Look into 'global dimming'.)

The next major consideration is whether or not the location is connected to the electric grid. Most of us have a connection to the electric power grid. If you are grid-connected, there are additional costs relative to the connection, but you will not need on-site energy storage (e.g., a battery bank). If you are off-grid, you need to produce all your own power. If you are grid-connected, you have the option of producing only a fraction of your own power, and if production and demand don't match up, well that becomes a problem for the grid managers.

There is one significant catch with many grid-connected PV systems that surprises many people. When there is a power outage, your battery-less, grid-connected household PV system may produce no output, even though the sun is shining. This is because many of the grid-connected PV systems rely on the presence of the grid to be their power sink. If it disappears, the system shuts down, even if there is demand within the household. So, if you are shopping for a grid-connected PV system, make sure it will provide you with power, even if the grid fails.

What do you hope to accomplish with the photovoltaic panels? What will the power be used for?

Are you planning to run an air conditioner? Look first into the potential for reducing solar gain, improving insulation, increasing thermal mass, and looking at geothermal potential or even solar-powered absorption or adsorption chillers for cooling before deciding to install a photovoltaic-powered air-conditioning system.

Are you planning to run an electric clothes dryer? If the sun is shining, you will be better off installing a clothesline or clothes tree and hanging the clothes to dry.

Are you thinking about lighting? If the sun is shining (to provide the photovoltaic electricity), could natural sunlight provide the desired lighting, via windows, skylights, or light tubes? If not, switching to high-efficiency lighting will be more cost effective than installing additional photovoltaic panels. Warning—basic math ahead.

Suppose you want to provide lighting in a basement 12 hours per day that is currently lighted by two 100-watt incandescent lights. 2 lights x 100 watts x 12 hours means a daily power consumption of 2400 watt-hours. Allowing for cloudy days and the fact that we get less than 3 full-sun equivalent hours on a good day in the winter in most of Ontario, let's presume we need to produce just 800 watts of power during daylight hours to provide a reliable 2400 watt-hours of electricity on a daily basis (optimistic). At current prices (approximately Cdn $7.50 per peak watt of PV capacity), this comes to Cdn $6000. And that doesn't include any costs for installation, grid connections, or batteries for local energy storage. And the PV panels will likely have to be replaced in 20 to 40 years.

Still want to install those PV panels? Let's consider some other options. Let's replace the incandescent lights with compact fluorescent lights (CFLs). 23-watt CFLs will produce slightly more light than the 100-watt incandescent bulbs. These lights cost about Cdn $5 each, for a total of Cdn $10. The lamps will likely last about 10,000 hours, or about 2 years in this application. 20 years worth will cost Cdn $100. If incandescent bulbs had been used, which lasted over 1000 hours each and cost Cdn $1.00 each, they would have cost Cdn $160 (4 bulbs per year per socket x 2 sockets x 20 years x $1/bulb). Now we need just 552 watt-hours per day to produce an equivalent amount of light (23 watts/lamp x 2 lamps x 12 hours/day). At Cdn $7.50 per peak watt, and still assuming an average of 3 full sun-equivalent hours per day, this comes to Cdn $1380 worth of PVs, plus the $60 saving for the CFLs over the incandescents for a total of Cdn $1480, a savings of Cdn $4520 relative to lighting the incandescents using photovoltaics.

Looking at it another way, would you rather spend $4620 to light incandescent bulbs with extra PVs, or save $60 on CFLs so all that electrical power is not required?

Assume (optimistically) a PV panel can be installed for Cdn $15 per peak watt (because the panels are typically about half—or less—of the cost of an installed system) and will last 20 years (at rated power), and produce rated power for an average of 2 hours a day (allowing for less than perfectly sunny days and lower-light periods in the morning and evening). The cost per kWh comes out to Cdn $1.27; $15 per peak watt x 1000 watt-hours/kWh divided by (2 hours/day x 365 days/year x 20 years = 14,600 hours). Still think it makes sense to invest in PV panels in the next few years? I know conservation (like upgrading insulation and weather sealing) and installing compact fluorescent lighting is not nearly as dramatic as installing solar (PV) panels, but it makes a lot more financial sense.

The fact is that most parts of Canada are not great candidates for PV panels (there are some notable exceptions). Take a look at a solar insolation map. Taking Ontario as an example, it will show that the entire province gets less than 3 kWh/sq. meter of effective sun on average per day. Some parts get less than 1. Toronto and southern Ontario get about 2. Ottawa and northern Ontario do better, at about 2.4 kWh/sq. meter per day. And that doesn't even consider smog and haze effects in urban areas that would diminish these figures further. Compare this to places like Nevada that get 7 kWh/sq. meter a day of solar insolation on average. Be sure you understand the insolation patterns in your location, including local shading, before shelling out for a PV system.

There is some mythology regarding the issue of energy payback from photovoltaic panels. Some suggest that the panels are a net energy sink, rather than an energy source, postulating that it takes more energy to make a PV panel than will ever be recuperated through its use. Piffle. There may have been some merit to this idea in the very early days of PV production, when the cells were typically castoffs from the solid-state electronics chip production factories. However, today, cells are being produced in quantity specifically for use in PV applications. Given PV panels are bought and sold in a free market, and a cost-effective solution to electrical generation in some applications, it should be intuitive that they have to be able to produce more energy during their life than it takes to produce them, or they wouldn't be economically attractive to purchase. This assumes they will be installed in a rational manner. Certainly, if a PV panel is stored in the dark for a century, it may well not produce as much energy as was used to produce it before it dies of old age, but such is hardly the typical case.

For those who desire a more scientific approach, there have been studies done that conclude PVs are net energy sources.[6][7][8][9]

---

6.    There is a good survey article on the subject titled: *Energy Leverage of Photovoltaics* at the ecotopia Web site.
      http://www.ecotopia.com/apollo2/pvlever.htm
7.    *PV Payback* by Karl E. Knapp and Theresa L. Jester
      Home Power Magazine Issue 80 (December 2000–January 2001)
      http://www.homepower.com/files/pvpayback.pdf
8.    *Can Solar Cells Ever Recapture the Energy Invested in Their Manufacture?* by Richard Corkish Solar Progress Journal (Australia and New Zealand Solar Energy Society) volume 18, Number 2, pages 16-17 (1997)
      http://www.ecotopia.com/apollo2/pvpayback.htm
9.    *The Energy Intensity of Photovoltaic Systems* by Andrew Blakers and Klaus Weber (Centre for Sustainable Energy Systems, Engineering Department, Australian National University, 2000)
      http://www.ecotopia.com/apollo2/pvepbtoz.htm

Perhaps the ultimate test for the net energy sink vs. net energy source question is whether or not a PV manufacturing plant can produce enough energy to keep producing more PV panels. That question was answered in 1982. In 1982 two PV-powered plants for the production of PV panels went into operation: the Solarex plant in Frederick, MD (200 kW) and the ARCO Solar plant in Hisperia, CA (1 MW). Both plants were capable of producing more energy from their solar power panels than they required to produce more panels.

I have not been able to find a credible article that suggests PV panels actually produce less energy over their life than is taken to manufacture them.

If you live in an area with good sun resources, have run out of other things to do to reduce your energy consumption, and feel the economics of photovoltaics justify the investment in your situation, then please don't let this myth stand in your way.

## 27.7. Your Path

Other books serve as guidebooks, with a paint-by-numbers approach to showing you what you can do. While I have provided some examples in this book of things you can do, that is not my basic objective. Instead, I hope this book serves as a framework and you will create your own solutions. I don't live in your circumstances. I don't know what you know. Why should anyone expect that I am going to solve your specific problems? I encourage you to do the following.

### Question

Much of what we think we know is not actually true. There is much I had to unlearn in order to write this book. If something strikes you as being incorrect, or inappropriate, question it, think about it, and research it.

### Learn

I hope this book has helped you. Don't stop here. There is a wealth of other information out there for you. The endnotes and bibliography in this book are just a couple of starting points.

### Practice

Knowledge has little value if it is not applied for your benefit. However small, try things that you expect will improve your situation. You will also learn from your experience.

## Share

When you know something works, share the knowledge. Lead by example. Teach others what you know to be true.

# 28

## *Conclusions*

For most people, the hydrogen economy is attractive because they hope it will provide copious amounts of affordable energy for future use, derived from sustainable and environmentally benign sources, while producing a minimum of noxious or undesirable by-products.

Environmental benefits notwithstanding, the real driver for renewed interest in the hydrogen economy is the growing realization that we are running out of cheap, easily accessible, conventional oil, probably within this century. Hydrogen fuel cell vehicles were proven to work in the 1960s. The key change in the succeeding four decades is not the technology, but the impending arrival of scarcity of conventional oil on a global scale. However, the critical point where demand exceeds supply is likely to arrive sooner, probably within the next generation. While still open to debate, it is my belief that the global Hubbert's Peak for production of conventional oil has already passed. Right now, we're delaying the inevitable by using reserves faster and supplementing them with small amounts of more expensive, unconventional sources.

Most consumers have no idea how dependent we are on oil and products derived from it. If our true dependency were known, consumers would consider our shrinking reserves too valuable to burn.

Some still deny that global climate change is real, let alone that it is happening now. Well, the Arctic ice cap has shrunk, measurably, in the past decade. Sections of the permafrost are thawing. Habitat areas for many species are changing. While specific climate effects are being observed on a local basis, noticeable changes are also occurring on a global basis. It will get worse before it gets better because there is a degree of inertia involved. This may be part of a normal cycle for the planet, but more likely it is being made more extreme by an increased inventory of greenhouse gases in our atmosphere, largely attributable to human activity (increased consumption of fossil hydrocarbons, reduced vegetation as a carbon store, increased numbers of ruminant livestock, etc.). Survival of the human species, let alone life as we know it, is not a given. I know that sounds alarmist. However, the alarms are sounding, even if we choose to ignore them. Some believe that the hydrogen economy is part of the solution to this situation, because less greenhouse gases would be produced as a result of using hydrogen instead of hydrocarbons as a dominant fuel.

The hydrogen economy is a bad idea. A really, really bad idea. Studies, even by the U.S. government, show the hydrogen economy will use more primary energy than the current technological environment. It will hasten the depletion of our existing oil and natural gas reserves. It will generate more greenhouse gases than our current practices. It will cost billions to trillions of dollars to implement. Even with a few miracles and a "moon-shot" program, it won't be ready in time to meet the impending shortages of oil and natural gas for the industrialized world.

We can make a technically feasible HFCEV—that's the easy part. What will be much harder is building a hydrogen infrastructure that is efficient, economical, and doesn't do more environmental damage than the status quo. The hydrogen economy will actually increase our greenhouse gas emissions. It will increase our use of primary energy, and the environmental damage associated with using more energy than we do today. Use of more energy means the hydrogen energy cycle will be more expensive than our current practices. If we want solutions we can afford, economically and environmentally, we have to look to things other than the hydrogen economy.

The amount of energy produced from sustainable sources today is trivial relative to the amount of energy we use. Anyone who does not accept that the fundamental first step to getting to the hydrogen economy (assuming they still believe in it) is to develop the sustainable energy sources that can be used to produce the hydrogen they desire is living in a fool's paradise, times two. This is certainly the case for governments and industry around the world today, as they prepare to implement the infrastructure for the distribution, storage, and use of hydrogen on an increased scale. At the same time, however, they are actually reducing funding, support, and implementation of additional sustainable energy sources.

Worse still, the envisioned hydrogen economy will actually increase demand for energy and waste heat production relative to today, other things being equal. This is the inevitable result of the additional conversion steps associated with producing hydrogen as the energy store, and then converting it again to obtain useful work. This means the hydrogen economy will also require us to reduce our fundamental demands for products and services that consume the energy we use today. Instead, we should embrace energy efficiency and conservation measures.

It is always cheaper to reduce energy consumption than it is to invent new ways to produce energy.

The hydrogen economy may work for Iceland. Iceland has access to significant quantities of sustainable geothermal energy, and no reserves of fossil fuels within its borders. Iceland is probably the best candidate on the planet for a viable hydrogen economy. However, even Icelanders would probably be better off with the electricity energy economy.

ABEVs are a better bet for the transportation sector than HFCEVs. ABEVs can be available for mass use sooner, will be less expensive to build, more efficient in terms of overall energy use, and safer. The logical interim steps to ABEVs include BEVs using current off-the-shelf battery technologies and hybrid vehicles. Hybrid vehicles must advance from their current state (electric-assist fossil-fuelled vehicles), to PHEVs which are powered by both electricity and fossil fuels, and finally to PHEVs powered by electricity and biofuels. This change alone will not be sufficient. We need to scale down the size of our vehicles, incorporate more efficient alternatives such as human-electric hybrids, electric-assist bicycles, and innovative vehicles like the Twike.[1] We also need to reduce our overall demand for transportation (in ton-miles, tonne-kilometres, passenger-miles or passenger-kilometres), to reduce the associated energy demand.

The current U.S. administration doesn't have a real energy or environment plan, so much as they have an oil industry aid plan. The Canadian federal government ceased having an energy plan for Canada when Liberal Prime Minister Jean Chrétien signed the NAFTA agreement after winning an election by campaigning against it. If citizens want a sane approach to energy or environmental policy, they will have to develop it themselves. Clearly neither the U.S., nor Canadian governments, nor the large global corporations have a long-term vision on energy let alone one that moves us to a sustainable energy production and use model. Governments, on the whole, are more problem than solution in addressing the issues associated with the transition from the hydrocarbon economy to a successive paradigm. Government actions in aggregate are an impediment to the rational changes that are required, that will happen sooner (more smoothly over a longer period) or later (more abruptly and with greater upheaval).

To say that waiting for a viable hydrogen economy is analogous to hoping for a miracle understates the case. It is more like depending on a series of miracles. The technology simply is not ready, let alone economic. Even if we believe the

---

1.    Twike human-electric hybrid car
     http://www.twike.ca

hydrogen economy is desirable and will come to pass, the winners in terms of competing technologies in various aspects of the hydrogen fuel cycle are not yet known. For example, installing hydrogen-fuelling stations is premature, given that the methanol fuel cycle is at least as feasible at this point, and the already ubiquitous electricity infrastructure is more attractive still. Even U.S. government agencies concede that fuel cells must improve by at least an order of magnitude in cost, double in efficiency, and last five times longer than they do today in order to become practical.[2]

Whether the hydrogen economy comes to pass or not, the necessary first step to it, increasing our supplies of sustainable energy, is a good idea. It is good for us environmentally, economically (at least at the macro scale), and also from a viewpoint of stability. The sun will be shining, the wind will be blowing, and the rain will continue to fall after we can no longer afford to use oil and other fossil fuels the way we do today and we have exhausted our supplies of economically recoverable uranium.

Unfortunately, our societal practice of subsidizing the price of oil and other fossil fuels is impeding us from embracing their replacement with the sustainable energy sources and energy conservation measures. We need to be putting these new measures into place now, in order to make the transition a smooth one. Moving to realistic pricing of energy resources in the short term will simply encourage us to start this journey sooner. In our current economic structure, it is difficult to profit by selling less of a commodity, a barrier we need to address to make a conservation culture viable.

There are those that believe if we invest enough in technology and R&D, we will find the answers that will make the hydrogen economy a success. I partially agree. We do need to invest significantly in technology and R&D. Not in hydrogen infrastructure, but in making sustainable energy technologies more usable, e.g., more affordable differential thermostats, solar collectors, automatic insulating window blinds, other diurnal cycle insulation for windows, batch solar water heaters, better solar cookers, advanced batteries for PHEVs, ABEVs, and other lighter-footprint zero-emissions transportation technologies.

---

2.   *The Hydrogen Economy: Opportunities, Costs, Barriers and R&D Needs*, 2004, U.S. National Research Council
http://books.nap.edu/books/0309091632/html/
Essentially, the NRC concludes that the DOE's timeline projections are aggressive, and that much fundamental research is still required. Progress to meet DOE targets will not come from refining existing technologies, but breakthroughs will be required to make the hydrogen economy a practical reality.

The good news is that reducing our dependence on environmentally destructive fossil fuel use (and the implicit production processes) is relatively easy, once individuals throw off their blinders and apathy, and choose to do something positive.

Making a positive change on the energy front is within the grasp of the average consumer. These changes will also help consumers save money, improve their quality of life, and the quality of our environment.

Neither (senior) governments nor major, multi-national corporations have a vested interest in making this easier for the majority of people on the planet. If we want a workable plan for the transition from cheap oil, we will have to develop it ourselves, as rational, self-interested individuals. Humans are more complex than economists can understand. Our satisfaction (utility to economists) is not dependent simply on material things. We also crave leisure, community, a degree of stability (but not stagnation), and other intangibles. Economic theory that sees people only as consumers of resources and providers of labour does not understand humans, and cannot hope to model them accurately. As living entities, and as a species, we need sustainable supplies of energy and practices to produce food and other goods, and a habitable planet for humans into the future.

It's your planet. If you won't look after it, who will?

# APPENDIX A

## *The Hydrogen Timeline—Historical*

| | |
|---|---|
| 1520 | Paracelsus discovers a gas that was likely hydrogen. |
| 1671 | Robert Boyle dissolves iron filings in dilute hydrochloric acid, and discovers the off-gas was highly flammable. |
| 1766 | Henry Cavendish demonstrates that burning hydrogen produced water. |
| 1766 | Antoine Lavoisier names the gas burned by Cavendish, "hydrogen" (for "that which makes water"). |
| 1812 | World's first city gas (coal gas, primarily hydrogen) plant constructed in London, England. By mid-1800s, city gas was in widespread use, with many cities having a city gas works, for street lighting, heating, and powering engines. The Dunedin Gas Works in New Zealand only ceased operation in 1987, and is now a museum. |
| 1837 | First (non-rechargeable) electric car ran under its own power. |
| 1843 | Sir William Grove developed the first working hydrogen fuel cell (pre-dates the rechargeable battery and the internal combustion engine). |
| 1859 | First rechargeable battery (not suitable for vehicle use). |
| 1861 | First Otto cycle engine. |
| 1881 | First rechargeable electric car in operation (Trouve, France). |

| | |
|---|---|
| 1892 | First Diesel engine. |
| 1898 | Sir James Dewar (Dewar Jar) produces liquid hydrogen. |
| 1900 | The first lighter-than-air craft lifted by hydrogen makes a successful flight. |
| 1925 | U.S. Helium Control Act enacted. |
| 1931 | Harold Urey discovers deuterium (heavy hydrogen). |
| 1932 | Francis Thomas Bacon develops his first Bacon cell, an alkaline fuel cell. Development continues for decades, and practical devices emerge in the late 1950s. |
| 1937 | On May 7, the *Hindenburg* zeppelin catches fire at Lakehurst, New Jersey, and sinks to the ground. 35 people die, 62 survive. |
| 1959 | Allis-Chalmers demonstrates first fuel cell vehicle, a D-12 tractor. The vehicle used propane as its fuel (not hydrogen), and an alkaline fuel cell (likely a Bacon cell). |
| 1964 | Roger E. Billings modifies a Model A Ford truck to run on hydrogen. Billings remains a leading advocate of hydrogen power to the present day. |
| 1965 | NASA first uses fuel cell in space with Gemini V. It was a problem with a fuel cell that causes the explosion on Apollo 13 during its trip to the moon. |
| 1967 | GM ElectroVan developed. This vehicle used a 5 kW fuel cell from Union Carbide, necessitating a battery/fuel cell hybrid power supply arrangement. The vehicle used liquid hydrogen storage. Range was approximately 200 km. The vehicle was never permitted to operate on public roads. |
| 1970 | Dr. Karl Kordesch builds a fuel cell / battery electric hybrid car for his own personal use. Storing hydrogen gas at up to 150 bar, the vehicle boasted a range of up to 300 km (180 miles). |
| 1996 | Metallic hydrogen created (at over 1 million bars pressure). |

| 2003 | On September 7th, California gubernatorial candidate Arnold Schwarzenegger promises to convert one of his Hummers to hydrogen power, whether he is elected or not. Although 4 bids are received to carry out the conversion, no contract is ever let, and the conversion is never done. |

| 2004 | On April 20th, Governor Arnold Schwarzenegger signed an executive order creating a partnership to build a network of hydrogen stations across California by 2010 at an estimated cost of US $100,000,000.00, for less than 200 stations in total (over US $500,000.00 each). |

| 2004 | On October 22nd, California Governor Arnold Schwarzenegger held a press conference at Los Angeles International Airport (LAX) to unveil his Hummer converted to hydrogen power (as committed to in 2003 election campaign), and the first public hydrogen filling station on the California Hydrogen Highway at LAX. |

Unfortunately, the Hummer H2 presented did not belong to the Governor. It was a prototype supplied by General Motors, boasting a range of only 50 miles per refill. Suddenly, the commitment to put a hydrogen filling station every 20 miles on the California Hydrogen Highway makes an eerie kind of sense.

The prototype hydrogen Hummer is so fragile, it is not permitted to be used off-road. To complete the farce, the hydrogen filling station used as a backdrop was not operational, is not open to the public, and will not be for years. As a result, the hydrogen Hummer had to be transported to and from the event, and not driven, other than the few yards it was driven by the Governor as part of the photo-op.

GM officials did not provide figures as to how much had been spent on the development and construction of the hydrogen-powered H2 prototype, which uses hydrogen in an internal combustion engine, and does not use fuel cells.

2004                        On December 2nd, U.S. bipartisan National Commis-
                           sion on Energy Policy releases a report titled *Ending the
                           Energy Stalemate: A Bipartisan Strategy to Meet America's
                           Energy Challenges*, that states, "hydrogen offers little to
                           no potential to improve oil security and reduce climate
                           change risks in the next twenty years".
                           http://www.energycommission.org/ewebeditpro/items/
                           O82F4682.pdf
                           This report was ignored by the mainstream media, and
                           nobody else paid any attention either.

2005                        The U.S. Army Corps of Engineers released document
                           ERDC/CERL TR-05-21, *Energy Trends and Their
                           Implications for U.S. Army Installations* by Donald F.
                           Fournier and Eileen T. Westervelt. It includes this
                           statement:
                           "In summary, there are tremendous technical hurdles to
                           overcome; once we have solved the production, trans-
                           mission and resource issues and then the switch to
                           hydrogen may occur. This is a long-term issue and the
                           hydrogen economy is decades away."

# APPENDIX B

## *Measures and Conversions*

All conversion figures are approximate.

## *POWER*

1 Watt = 1 joule per second
1 kiloWatt (kW) = 1000 Watts = 1.34 horsepower (hp)
1 horsepower (hp) (English) = 746 Watts

## *ENERGY*

1 kiloWatt-hour (kWh) = 3,600,000 joules = 859,845.23 calories =
1.34 horsepower-hours = 3,412 British Thermal Units (BTUs)
1 joule = 1 newton-metre = 1 Watt-second = 0.00095 BTUs

## *PRESSURE*

1 bar = 14.5 pounds per square inch (psi)
1 bar = 0.99 atm (atmospheres)
1 bar = 100,000 Pascals
1 bar = 100 kilopascals

## *TEMPERATURE*

The Kelvin temperature scale starts from absolute zero as its zero point, and uses
the same graduations as Celsius degrees.

To convert Kelvin to degrees Celsius, subtract 273.

E.g., the boiling point of water is 373 K. To convert to Celsius, subtract 273 (373–273 = 100), resulting in 100°C.

To convert from Celsius to Kelvin, add 273.

E.g., the freezing point of water is 0°C. To convert to Kelvin, add 273 (0 + 273 = 273), resulting in 273 K.

To convert from degrees Celsius to degrees Fahrenheit:
multiply by 1.8, then add 32.

E.g., the boiling point of water is 100°C, multiply by 1.8 to get 180, then add 32, to arrive at the result 212°F.

To convert from Fahrenheit to Celsius:
subtract 32, then divide by 1.8.

E.g., normal human body temperature is 98.6°F, subtract 32 to get 66.6, then divide by 1.8 to get 37°C.

## MASS

1 gram = 0.0353 ounces (oz)
1 kilogram (kg) = 1,000 grams = 2.2 pounds (lb)
1 metric tonne = 1,000 kg = 2,204.6 pounds = 1.1 tons
1 ounce (oz) = 28.35 grams
1 pound (lb) = 0.454 kg
1 stone = 14 pounds (lb)
1 ton = 0.907 (metric) tonnes

## VOLUME

1 cubic centimetre (cc) = 1 millilitre = 0.03 fluid ounces (U.S.) = 0.04 fluid ounces (Imperial)
1 litre = 1,000 cc = 0.26 gallons (U.S.) = 0.22 gallons (Imperial)
1 cubic metre = 1,000 litres
1 gallon (U.S.) = 0.83 gallons (Imperial) = 3.79 litres

# DISTANCE

1 millimetre (mm) = 0.0394 inches
1 centimetre (cm) = 0.394 inches
1 metre = 3.281 feet
1 metre = 1.094 yards
1 kilometre (km) = 0.6214 miles
1 inch = 25.4 millimetres (mm) = 2.54 centimetres (cm)
1 foot = 0.3048 metres
1 yard = 0.9144 metres
1 mile = 1.609 kilometres (km)

# VELOCITY (SPEED)

1 kilometre per hour (km/h) = 0.62 miles per hour (mph)

If these conversions don't suit your needs, there are many good converters available on the Internet now, so you can convert those confusing metric (Systeme Internationale) units like metres/second and km/h into infinitely more useful American (ironically, formerly British Imperial) measures like furlongs per fortnight or chains per century.

Note to sci-fi screenwriters: a light-year is a measure of distance, not time. It's about 9,460,536,207,068,016 metres, or about 5,878,502,843,522 miles. It's just really far, OK? Oh, and while we're on the subject, a parsec (parallax-second), is also a measure of distance, and not time. It's about 3.26 light-years, or really, really far.

# Appendix C

## *Glossary*

| | |
|---|---|
| ABEV | Advanced battery electric vehicle |
| AC | Alternating current |
| A/C | Air conditioning |
| Anaerobic | Without oxygen |
| BEV | Battery electric vehicle |
| CAFE | Corporate Average Fuel Economy |
| CARB | California Air Resources Board |
| CFL | Compact fluorescent light |
| CHP | Combined heat and power |
| CNG | Compressed natural gas |
| Co-generation | 1) Producing heat and power from same installation, synonymous with combined heat and power (CHP)<br>2) Small electrical generator connected to the grid to provide power to other utility subscribers (less common definition today than 1) above. |
| CoP | Co-efficient of performance |
| DC | Direct current |
| DG | Distributed generation |
| DOE | (United States) Department of Energy |
| DSM | Demand side management |
| EPRI | Electric Power Research Institute |

| | |
|---|---|
| EV | Electric vehicle |
| GHG | Greenhouse gas(es) |
| Heliostat | A reflective surface, usually movable, aimed to reflect sunlight onto a focal point to increase the energy received at that point. |
| HFCV | Hydrogen fuel cell vehicle |
| HFCEV | Hydrogen fuel cell electric vehicle |
| HICEV | Hydrogen internal combustion engine vehicle |
| ICE | Internal combustion engine |
| Insolation | Energy from the sun received at a given point, e.g., the Earth's surface. |
| kW | Kilowatt (measure of power) |
| kWh | Kilowatt-hour (measure of energy) |
| Kyoto | Shorthand for the Kyoto Protocol, the United Nations agreement on curbing greenhouse gas emissions |
| LPG | Liquid propane gas or liquid petroleum gas |
| LSV | Low speed vehicle<br>The Canadian legislative equivalent of an NEV in the U.S.—an electric drive car (Canada excludes trucks from its regulations) governed to a maximum speed of 40 km/h. |
| MOX | Mixed OXide (fuel for nuclear reactors, usually plutonium and depleted uranium) |
| MW | Megawatt—one million watts or one thousand kilowatts |
| NEV | Neighbourhood electric vehicle<br>A category of vehicles powered by electricity and limited to a maximum speed of 25 mph. |
| NG | Natural gas—a fossil fuel, primarily methane |
| NHTSA | (U.S.) National Highway Traffic Safety Administration |

| PEM | Proton exchange membrane (also polymer electrolyte membrane, less common) |
| PNGV | Partnership for a New Generation of Vehicles<br>A U.S. government sponsored program in the 1990s focused on the development of a high mileage vehicle (80 mpg gasoline equivalent), notably diesel-electric hybrids. |
| PR | Public relations |
| PSI | Pounds per square inch (usually written in lower case, i.e., "psi") |
| PV | Photovoltaic<br>A device that produces DC electricity from light. |
| R&D | Research and development |
| SAE | Society of Automotive Engineers (U.S.) |
| USABC | United States Advanced Battery Consortium<br>A collective effort by U.S. automakers sanctioned by the U.S. government in the 1980s to develop advanced batteries for electric vehicles. |
| USDOE | United States Department of Energy |
| V2G | Vehicle to grid<br>Using vehicles (e.g., hybrids or hydrogen fuel cell vehicles) to produce electricity to be fed into the grid. |
| ZEH | Zero energy house<br>A house designed to produce energy, such that net energy consumption for the household is nil over an extended period of time. |
| ZEV | Zero emission vehicle<br>A vehicle (typically on-road) that produces no measurable pollutants or noxious emissions in operation. |

# *Bibliography*

Despite its size, this bibliography is incomplete. I began my researches in some areas covered in this book long before I knew I would be writing it. So some sources, undoubtedly important ones in shaping my thinking, have probably been missed as I have reviewed my notes and racked my brain trying to remember where various bits of knowledge I have acquired were originally brought to my attention. For that, I apologize in advance. I hope that readers will find this list sufficient to guide them in further studies, if they are so inclined. The book covers a lot of areas, so this list should lead to many interesting places.

Arlington Institute, The, *A Strategy: Moving America Away From Oil*, August 2003
http://www.arlingtoninstitute.org/library/A%20Strategy%20-%20Moving%20America%20Away%20from%20Oil.pdf

Bakan, Joel, *The Corporation: The Pathological Pursuit of Profit and Power*, Penguin Canada, 2004, ISBN 0670889768

Bradsher, Keith, *High and Mighty: SUVs—the world's most dangerous vehicles and how they got that way*, New York, Public Affairs, 2002, ISBN 1586482033

Carson, Rachel, *Silent Spring*, New York, Facet World Library, 1962

Coleman, Elliot, *Four Season Harvest*, New Chelsea Green, 1999, ISBN 1890132276

Davidson, Joel, *The New Solar Electric Home*, Ann Arbor Michigan, aatec publications, 1987, 1995, ISBN 0937948098

De Villiers, Marq, *Water*, Toronto, Stoddart Publishing, 2000, ISBN 0773761748

DeWarrd, E.John, and Klein, Aaron E., *Electric Cars*, Garden City, NY, Doubleday & Co., 1977, ISBN 038508143X

Doyle, Jack, *Taken For a Ride: Detroit's Big Three and the Politics of Pollution*, New York, Four Walls Eight Windows, 2000, ISBN 1568581475

Eaves, Stephen and Eaves, James, *Cost Comparison of Fuel-Cell and Battery Electric Vehicles*, Current Events magazine, September-October 2004 issue.

Greene, David L. and Santini, Danilo J., *Transportation and Global Climate Change*, Washington D.C., American Council for an Energy Efficient Economy, 1993, ISBN 0918249171

Goodstein, David, *Out of Gas—The End of the Age of Oil*, New York, W.W.Norton & Company, 2004, ISBN 0393058573

Gordon, David (editor), *Green Cities—Ecologically Sound Approaches to Urban Space*, Montreal, Black Rose Books, 1990, ISBN 0921689543

Heinberg, Richard, *Powerdown*, Gabriola Island, New Society Publishers, 2004, ISBN 0865715106

Heintzman, Andrew and Solomon, Evan (editors), *Fueling the Future: How the Battle Over Energy is Changing Everything*, Toronto, House of Anansi Press, 2003, ISBN 0887846955

Himmelman, William A., *Solar Engineering for Domestic Buildings*, New York, Marcel Dekker, Inc., 1980, ISBN 0824769902

Hoffman, Peter, *The Forever Fuel*, Boulder, Colorado, Westview Press, 1981

Ingrassia, Paul and White, Joseph B., *Comeback: The Fall & Rise of the American Automobile Industry*, New York, Simon & Schuster, 1994, ISBN 0671792148

Jones, Ellis, Haenfler, Ross, Johnson, Brett and Klocke, Brian, *The Better World Handbook*, Gabriola Island, B.C., New Society Publishers, 2001, ISBN 0865714428

Kirsch, David A., *The Electric Vehicle and the Burden of History*, New Brunswick NJ, Rutgers University Press, 2000, ISBN 0813528097

Klein, Naomi, *No Logo*, Vintage Canada, 2000, ISBN 0676972829

Koppel, Tom, *Powering the Future: the Ballard Fuel Cell and the Race to Change the World*, Toronto, John Wiley & Sons Canada, 1999, ISBN 0471644218

Korchinski, William J., *Fueling America: How Hydrogen Cars Affect the Environment*, Reason Foundation, November 2004

Korten, David, *When Corporations Rule the World*, Berrett-Koehler, 2001, ISBN 1887208046

Lomborg, Bjorn, *The Skeptical Environmentalist*, Cambridge University Press, 2001, ISBN 0521010683

MacKenzie, James J., *The Keys to the Car: Electric and Hydrogen Vehicles for the 21$^{st}$ Century*, World Resources Institute, 1995, ISBN 0915825937

Moore, Michael *Dude, Where's My Country?*, New York, Warner Books, 2003, ISBN 0446532231

Park, Jack, *The Wind Power Book*, PaloAlto, California, Cheshire Books, 1981, ISBN 091735205X

Perkel, Colin N., *Well of Lies—The Walkerton Water Tragedy*, Toronto Ontario, McClelland & Stewart Ltd., 2002, ISBN 0771070195

Radabaugh, Joseph, *Heaven's Flame—A Guide to Solar Cookers*, Ashland Oregon, Home Power Publishing, 1998, ISBN 0962958824

Rifkin, Jeremy, *The Hydrogen Economy: The Creation of the World-Wide Energy Web and the Redistribution of Power on Earth*, New York, J.P. Tarcher/Putnam, 2002, ISBN 1585421936

Romm, Joseph J., *The Hype About Hydrogen: Fact and Fiction in the Race to Save the Climate*, Washington, Island Press, 2004, ISBN 155963703X

Safdie, Moshe and Kohn, Wendy, *The City After the Automobile*, Toronto, Stoddart, 1997, ISBN 0773729836

Schiffer, Michael Brian, *Taking Charge: The Electric Automobile in America*, Washington D.C., Smithsonian Institution Press, 1994, ISBN 1560983558

Schnayerson, Michael, *The Car That Could*, New York, Random House, 1996, ISBN 067942105X

Sherman, Joe, *Charging Ahead*, New York, Oxford University Press, 1998, ISBN 0195094794

Sperling, Daniel, *Future Drive: Electric Vehicles and Sustainable Transportation*, Washington D.C., Island Press, 1995, ISBN 155963328X

Standing Committee to Review the Research Program of the Partnership for a New Generation of Vehicles, *Review of the Research Program of the Partnership for a New Generation of Vehicles*, Washington D.C., National Academy Press, 1994

Tataryn, Lloyd, *Dying For a Living—The Politics of Industrial Death*, Deneau & Greenberg, 1979, ISBN 088879018X

Vale, Brenda and Robert, *The Autonomous House: Design and Planning for Self-Sufficiency*, New York, Universe Books, 1975, ISBN 0876639791

Wakefield, Ernest H. *History of the Electric Automobile Battery-Only Powered Cars*, Warrendale PA, SAE Inc., 1994, ISBN 1560912995

Wakefield, Ernest H. *History of the Electric Automobile Hybrid Electric Vehicles*, Warrendale PA, SAE Inc., 1998, ISBN 0768001250

Wang, Michael, *Fuel Choices for Fuel-Cell Vehicles: Well-to-Wheels Energy and Emission Impacts*, Center for Transportation Research, Argonne National Laboratory, 2002
http://www.transportation.anl.gov/pdfs/TA/260.pdf

Wright, Patrick J., *On a Clear Day You Can See General Motors*, New York, Avon Books, 1979, ISBN 0380517221

# Index

# Notes

# *Notes*

# Notes

*Notes*

# *Notes*

*Notes*

# Notes

*Notes*

# *About the Author*

Author with his green electric car

After graduating with an Honours degree in Commerce (business administration), Darryl McMahon worked in the nuclear energy sector, and then in the software development business as a business and systems analyst, programmer, database designer, application architect, team leader, and project manager. He has published articles in the field of personal computing.

He has also been noted for his work in the area of sustainable transportation technology, notably electric and hybrid vehicles dating back to the 1970s. One of the original founders of the Electric Vehicle Association of Canada, he has published a number of articles related to electric and hybrid vehicles. He was a speaker at the international Electric Vehicle Symposium 17 in October 2000. His personal vehicle fleet currently includes electric cars, electric motorcycles, an electric assist bicycle, solar-electric boat, an electric garden tractor, and a truck that runs on biodiesel fuel, and he has more projects in the works.

He is the founder and President of Econogics, Inc. (its name is a blend of economics, logic, and ecologic). The company website (www.econogics.com) provides a wealth of information on consumer-level energy conservation and efficiency techniques, zero-emissions electric drive technology, and related subjects.

More recently, he has devoted some time to the economics of substituting sustainable energy sources (he uses the term "preferred energy") for conventional (fossil) energy sources, and to this end has built a solar water pre-heater, an active solar space heating system, solar cookers, and greenhouses—installing them all in a suburban setting.

An avowed curmudgeon, avid organic gardener, and Freecycler, he lives in Ottawa with his wife and son who tolerate (and occasionally encourage) his eclectic range of activities and projects, including taking training to become a certified photovoltaic installer.

978-0-595-39229-2
0-595-39229-6

www.ingramcontent.com/pod-product-compliance
Lightning Source LLC
Chambersburg PA
CBHW020728180526
45163CB00001B/149